Development of Computer-Based
Production Systems

Development of Computer-Based Production Systems

A. K. Kochhar, B.Sc., Ph.D.

Schools of Mechanical and Manufacturing Systems Engineering and
Manufacturing Systems Design Centre
University of Bradford

Edward Arnold

© A. K. Kochhar 1979
First Published 1979 by
Edward Arnold (Publishers) Ltd
41 Bedford Square
London WC1B 3DQ

Kochhar, A K
  Development of computer-based production systems.
  1. Production management–Data processing
  I. Title
  658.5′0028′54044   TS155

ISBN 0–7131–3404–6

Preface

The advent of the general purpose digital computer has had a large impact on the lives of ordinary people in the developed western world; perhaps more than any other device invented during the last thirty years. In spite of its very wide use for all types of applications, the computer still remains a mystery to most people. Even in advanced industries, employing large numbers of highly educated people, the computer is often treated by most managers as a tool to be shunned. They may make occasional use of the information produced by the computer, but very rarely get involved in the specification of the type of information processing system which should be used by the company.

Many manufacturing companies have invested large sums of money and manpower resources in their data processing functions. In a large majority of these organizations the use of computers is often restricted to routine applications such as accountancy and pay-roll calculations. Some companies have sucessfully developed systems for planning and controlling the flow of work on the shop-floor. Production control is the complex centre of manufacturing systems, involving large volumes of data transactions. There is also a need for a fast and accurate response to events in order to improve the quality of decision making. The computer is particularly suited to coping with such dynamical systems and will yield the most significant benefits, in the form of improvements in productivity and better utilization of capital resources, when employed in this area.

The price of computer hardware has fallen steadily, over the past few years, in relation to capacity and the range of features offered. A single large scale integrated (LSI) chip is more powerful than the big computers of earlier days. The appearance on the scene of powerful mini- and micro-computers is a manifestation of increasing availability of computing power at relatively low cost. As a consequence many companies are contemplating the use of computers to improve control of production processes. Progress is held back partly by production staff who lack understanding of the potential of computers in this area, and also by senior management awareness of the accounts of unsatisfactory computer installations, which permeate the literature. Very often, after the decision to use a computer for production control work has been made, the task of system development is left to computer professionals who rarely have any knowledge of day-to-day production processes and do not appreciate the problems of the users. The users do not get involved because they are seldom familiar with computer technology and system development techniques. It is a well documented fact that where users do not participate in system development, the result, at best, is a limited success. The best systems have been developed by companies in which senior officers have taken the leadership in managing the project. If managers can be made aware of the capabilities, limitations and pitfalls inherent in changing to computer-based production systems, they can exercise a dominant role in relation to computer specialists, and ensure that the computer system is designed to satisfy user requirements. While detailed knowledge of the intricacies of computer systems is not essential, the managers must acquire sufficient familiarity with the basic techniques used in the development of systems.

The products manufactured and marketed by a company undergo changes over a period of time. As a result it is frequently necessary to modify the system used to plan and control the products. The decision to use one particular planning or control procedure for all types of products, irrespective of the type of technology involved, can often lead to disastrous consequences. Frequently two companies engaged in the manufacture of the same or very similar products have widely different financial fortunes. In the case of the company which goes out of business the blame does not always lie with the product; it could be due to the manufacturing technology employed, and the systems and control procedures used by the company. If a company is to survive then its systems must be modified, as and when the need arises, to take account of changes. For computer-based systems, the necessary modifications can be made in time only if the managers are familiar with the details of the system currently in operation and the effort required to modify it. It therefore follows that to perform this task effectively the functional managers will have to take up additional responsibilities. The manufacturing manager, for example, must look after the operation of computer systems used in his department in addition to performing his routine duties. In any event the introduction of an effective computer-based production system should relieve him of some of his normal work load.

This book, addressed to all the people who are in any way concerned with the development of production systems, attempts to help readers interested in the design of computer-based production planning and control systems suitable for companies which make use of batch production techniques. Having read and understood the book they should be able to adapt the guidelines contained in it for making the basic decisions necessary to develop and implement a computer system for resolving the production problems of their company. The text should also be of considerable interest to undergraduate and postgraduate students on engineering, management and computer science courses. Many of these students are familiar with programming, while others can design compilers, diagnostic routines or efficient algorithms; very few of them, however, appreciate the problems of manufacturing companies and the potential role of the computer in the resolution of these problems. A thorough study of this book will enable readers to make a substantial and worthwhile contribution to the development of effective computer-based production systems.

The book is divided into a number of parts.

Chapter 1 is essentially an overview of the development of computer-based production systems. The areas explored in the main body of the text are summarized.

Part I of the book considers, in some detail, the important features of computer hardware and software.

The individual computer applications in the field of production systems, viz. production database, bill of materials processing, forecasting, customer order processing, production schedule creation, material requirements planning, capacity planning, order release and operations scheduling, work-in-process control and inventory control, are discussed in Part II of this book. The use of computer programs for testing values of parameters and decision rules is examined.

Part III deals with the management of the development of computer-based production systems. The successive stages of a typical computer project are described. This section seeks to guide users toward the implementation of an effective computer system.

Wide-ranging discussions with a large number of people in industry as well as the academic world have helped in the preparation of this manuscript. In particular I am obliged to Professor J. Parnaby who initiated my interest and work in this field and with whom I have had numerous useful discussions. I am also grateful to the late Mr Alan Hammond and to Dr R. A. Worth for reading through the entire script and for their many valuable suggestions which have resulted in marked improvements to the contents of this book. Thanks are also due to Mrs Nicholeen Cooper, Mrs Mavis Maltas and Mrs Glynis Wilkinson for their careful typing and retyping of the manuscript. I am privileged to claim the mistakes, such as there are, as entirely my own.

Bradford   A. K. Kochhar
1978

Contents

PART III INSTALLATION OF COMPUTER SYSTEMS

1
Computer-Based Production Planning and Control Systems— an Overview

Introduction

The production function occupies an important role in the activities of a typical manufacturing concern. The production problems faced by most manufacturing companies are very similar. The management is responsible for planning and controlling the flow of work through the production areas so that the orders received from the customers can be satisfied on due dates. During periods of reduced economic activity the management has to ensure that the production workers are engaged in useful work until the time at which further definite orders are received. It may be necessary to authorize the production of commonly used parts during such slack periods so that good use is made of available machines and manpower, and it is also possible to satisfy customer demand quickly when an upturn in economic activity takes place. The requirements for products which can be directly ordered by the customers are represented on a planning document frequently referred to as a master production schedule. This master schedule is based on definite customer orders as well as forecasts of future demand. All the future production activities depend upon the contents of this document. The production planning and control manager requires information about the components assembled to produce the end items, the materials needed for manufacturing these components and the quantities of these required materials along with the associated due dates. He has to plan the production work taking into account the alternative manufacturing routes, the machine and skilled manpower capacity required to produce the components and assemble them into the final product, and the existing load on the manufacturing facilities. Based on this and relevant material availability information, the best time to release a production order to the shop-floor has to be decided.

If the materials and manufacturing capacity required to carry out the necessary work are not available, then the release of an order only adds to the congestion on the shop-floor. The movements of the individual production orders on the shop-floor have to be monitored to ensure that they are making satisfactory progress, and that the orders can be delivered on time. In many industries, the individual production items are subjected to inspection after the completion of each or a key manufacturing operation. If it becomes necessary to scrap an item due to its unsatisfactory quality and the available quantity falls below the finally required quantity, then a new batch of work has to be planned for and launched at the earliest possible opportunity, so that the assembly operation and the subsequent delivery of the product is not held up due to the lack of one or more components. Fig. 1.1 is a simplified representation of the sequence of activities included within the production function of a typical manufacturing company.

Most manufacturing companies also supply service/spare parts to their

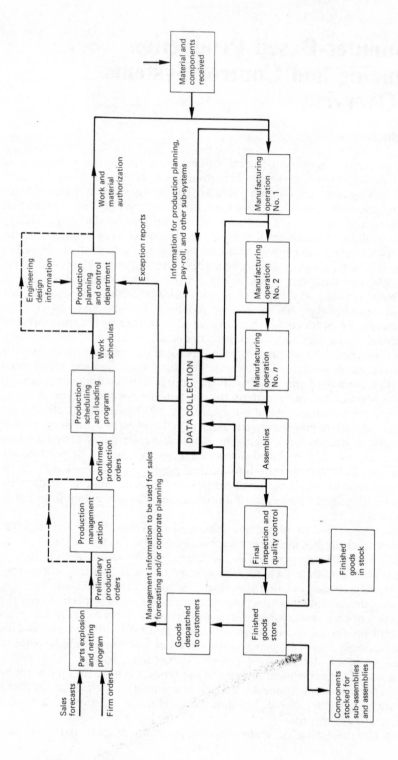

Fig. 1.1. The 'flow' of a typical manufacturing system and the associated information channels

customers. A high level of customer service may be attained by maintaining an inventory control system which can satisfy a majority of such requirements off the shelf. However, a balance has to be struck between the level of customer service, the amount of inventory carried by the company, and the efficient utilization of the manufacturing capacity. An efficient inventory control system is also a prerequisite for ensuring the availability of materials required to start production work. The material/component availability also depends upon the effectiveness of the purchasing function. Many manufacturing companies buy a high percentage of the components used in assembly operations. The required items can be obtained from a number of competing suppliers and the product quality and delivery performance factors are as important as the price of the item. An inefficient or badly managed purchasing operation can reduce the overall efficiency of the manufacturing system.

The production environment changes rapidly with time due to the large number of events which often take place at random intervals. The absence of workers, breakdown of machines, or lack of required materials can lead to a severe disruption of the planned work schedule. The ability of the management to plan and control the flow of work accurately is critically dependent upon the availability of timely, relevant, and accurate information. The people who make the decisions very rarely create the data or information on which these decisions are based. The data are continuously created in a number of departments and have to be stored for a period of time until updated by subsequent events. Alternatively these data items may lose their usefulness after a period of time in which case the appropriate records are deleted or destroyed.

A large volume of detailed and irrelevant information which cannot be comprehended by the busy line managers is of very little value to them. The flow of information within a department and between different departments defines the extent to which the activities of a manufacturing organization can be effectively planned and controlled. A good information flow system can make a substantial contribution towards the achievement of the primary objectives of a company. Such information flow systems have necessarily to cross the artificial departmental boundaries, and it is essential that the overall objective viewpoint should be considered during the design of these systems.

Manual Planning and Control Systems

The planning and control of production in even a medium sized company engaged in the manufacture of relatively complex products, requires the availability of a large amount of timely information. With manual systems such information is often too late, in addition to being inaccurate. When the required data does become available it is difficult to carry out the analysis necessary for extracting the relevant information, due to the lack of adequate time and resources. Under such conditions it is difficult to detect the underlying reasons for the variations in performance.

Frequently the planning clerks spend days preparing the work schedules to be used in the following weeks, only to find that the requirements have changed, and quick alterations have to be made to reflect the new circumstances. With a manual system it is impossible to show accurately the current and planned activity state of the company. Accurate planning based on a realistic appreciation of the relevant timely information reduces the need for very close control and the management does not have to resort to frequent expediting of critical items.

With conventional manual systems it is difficult to exercise adequate control over production work without employing a large number of progress chasers and clerks. Even assuming that it is economically feasible to employ such large numbers of people, it would still be difficult to prepare all the up to date reports required for management attention and action.

With manual stystems a frequently encountered problem is the lack of communications between different departments. The activities of a large number of departments with different objectives, and whose managers do not always see eye to eye, have to be co-ordinated to achieve an effective production control system. Production engineering, purchasing, design, sales, accounts, quality control, production shops, all interact with the production planning and control department and require the use of common data items. But due to the communications problems encountered in most manufacturing companies, decisions are frequently based on erroneous data values. Independent records are used in different departments of the company thereby increasing the clerical effort required to maintain what are often inaccurate and inconsistent files. These records are subject to continuous changes which have to be communicated to all the affected areas. This results in a vast amount of paperwork circulating between different departments. Often the departments in question are not informed about these changes or there is considerable delay in the paperwork reaching the people concerned. As a result the decisions made during the intervening period are based on the use of obsolete data.

It is difficult to conceive of the use of a manual centralized records system in the dynamic production environment. The constant stream of people waiting to retrieve or update these central records would render such a system totally impractical and the different company departments would start maintaining informal parallel records.

Computer Applications In Production Planning and Control

The essential role of the computer in the production function is to capture and process the data relating to a large number of transactions which continuously take place in the different departments of the company. The processed data, in the form of timely summary reports, and exception reports which highlight the deviations from planned operations or expected behaviour, can be used for management decision making. Computers by themselves do not make important decisions, although routine decisions can easily be automated. All the important decisions still have to be made by the management. The computer is a tool which can be used to control production effectively and provide assistance to the managers, faced with a series of constant changes, in the task of making necessary decisions. However, the computer is only a small, although very important part of a manufacturing information and control system. Information and control cannot be separated due to the fact that the information sub-system is a component of the overall control system.

or fortnight in advance, quickly become out of date due to the random events on the shop-floor whose effects multiply very rapidly. Such systems quickly fall into disrepute and do not result in any improvements. Frequently they have the opposite effect, i.e. the existing bad conditions get worse because user requirements cannot be taken into account in the system development process. A poorly designed and badly implemented computerized production planning and control system is worse than a good manual system which is thoroughly understood by all the employees concerned.

The packages currently available are better in many respects. It is possible to customize some of these packages by using available program exits. Often, however, the effort required for a detailed study of the system manuals and the amendments which have to be made, can be more than that required for producing a new system.

The introduction of a computer into the manufacturing area by itself does not result in better management of resources or higher profits. Similarly the use of computers does not lead to improved control of inventories. Computers can only assist in providing the relevant information. Best results are obtained by ensuring that the concepts used for planning and controlling the production are sound and fully understood by the people concerned. The key to success lies in the ability of the management to appreciate the problems in the production area and use the computer effectively for carrying out required data manipulations. Considerable improvements are possible by building additional control procedures into the programs used to carry out the required data processing. The responsibility for designing such procedures lies with the senior production management, and not with the data processing specialists. An important consideration in the implementation of a computer system is the design of supporting manual and clerical procedures. It is often necessary to alter the existing organization or departmental structure. Only the senior company management can make such decisions. Therefore the functional management must play a leading role in the task of system development.

The management has, of necessity, to delegate the responsibility for detailed system development work. However it must continuously monitor the progress of such work and, if necessary, exert influence to ensure that the overall business objectives and fundamental requirements are served. The computer application has to be designed so that it helps the operating personnel in carrying out their day-to-day duties.

The first stage in the development of an effective computer-aided production control system is the accurate identification of the information required to plan and control the production of goods. The 'systems approach' should be used in the development of these overall objectives. While difficult to comprehend fully the systems concept offers considerable benefits. Although the departmental managers have to pay particular attention to their own day-to-day problems, they must not lose sight of the complex interactions between the large number of factors that have an effect on the efficient operation of the production function.

User involvement in all aspects of system development work is a primary requirement for ensuring the success of any computer application. A departmental manager who is not fully convinced of the desirability of computerizing a particular function will hold back his support, and the resulting system even at best can only be a limited success. The users are familiar with their information requirements and can make a substantial contribution to the development and implementation of an effective computer-based system, provided they appreciate the potential use of computers in their daily working environment.

If the managers of the departments whose functions are being computerized take a leading role in the specification and development of the system, then the resistance on the part of the users to the implementation of such a system is minimized. The managers can also help educate the users in the department about the computer-based system. Employees are likely to notice the statements made by their managers who can also dispel any of the fears about the system. Outside data processing experts are often distrusted and resented by many employees. All employees should be fully informed about the workings of the overall system, and the vital role which they have in ensuring its success. If they are familiar with the mechanics of the total system, in addition to their own role, then they can also make suggestions for possible improvements.

A successful control system must have the basic property that it is at least as dynamic as the activities it is trying to control. The production control environment is extremely dynamic and requires the use of an on-line real-time computer system in which the transactions data are captured at source, and the affected files immediately updated. Such systems are almost always up to date. The only time delay is in the reporting of shop-floor movements data. There will always be a time delay between the actual completion of an operation and the input of relevant data to the system, unless the production machines are directly connected to the computer. While real-time systems are more complex and expensive than the equivalent batch processed systems, they do provide the timely and relevant information required for making decisions.

The format in which the information is presented to the users is a factor of considerable importance. Several otherwise well conceived systems have failed due to the fact that the data processing results are in the form of lengthy computer print outs. The busy manager has to spend a long time searching for potentially useful information which is dispersed throughout largely irrelevant data. Too many data items presented at the same time require considerable thinking on the part of individual users, and do not result in a practical system which is acceptable to users at all levels. The use of summary and exception reports, complemented by the application of terminals such as visual display units or teletypewriters which can be used to retrieve detailed information, makes it possible to develop really effective systems with which all the users can easily identify.

The development and implementation of a computer system takes a comparatively long time and requires substantial capital and manpower resources. It is usually assumed that the system under consideration will be used for a long period of time. Most manual systems involve the use of informal and ad hoc procedures. The use of such an approach with a computer-based system can only lead to disastrous results. A computer-based system must evolve as a result of detailed discussions between the senior managers of all the departments who will either input data to the system or receive reports produced by the processing of this data. It is important that these senior managers should have detailed knowledge of the role which the computer could play in the improvement of the working of their departments. If the managers do not appreciate the potential of the computer, then the computer-based system will simply automate the existing manual procedures, and the opportunity to streamline these procedures at the time of introduction of the new system will be lost.

The implementation of an effective computer-based production system can show substantial benefits. The major advantages are as follows.

1 Improvement in overall company profits.
2 Better machine and manpower utilization.
3 Considerable savings resulting from reduction in inventory levels of work-in-progress, raw materials, and finished goods.
4 Reduction in purchasing costs. Even small reductions when spread over a large number of items can have an appreciable effect on profits.
5 Speedier availability of better quality information. As a result decisions are based on up to date information.
6 Ability to expand business without proportional increases in staff.
7 Improvements in customer service level by making the manufacturing unit more effective. This enables the company to be more competitive and achieve an increased share of the market.
8 Ability to compare alternative and feasible solutions following digestion of all the information.
9 Compatibility of information used by different departments.
10 Selective reporting on shortcomings such as delayed deliveries or jobs falling behind schedule.
11 Facility for recording cost variances as they occur and reflecting them in selling prices very quickly.
12 The ability to ascertain machine, labour, and material requirements makes it possible to identify quickly areas which require attention.
13 The possibility of using better planning techniques by incorporating them into the overall computer system.
14 Systematic execution of company policy through the use of data handling and planning rules. The managers have time to make real decisions instead of filling in forms. This also gives them a better understanding of how to run the business.

The advantages quoted above depend on the effectiveness of the systems designed, the techniques used and the activities on which computers have made an impact.

We will now consider in detail the important features of (a) the computer technology required to implement effective computer based production systems; (b) the development of system modules for individual applications such as bill of materials processing, requirements planning, inventory control; (c) the design of a typical computer system, and the procedures used to install a successful system.

A familiarity with the subject matter of these topics will enable engineers and managers to take a leading role in the development of production systems.

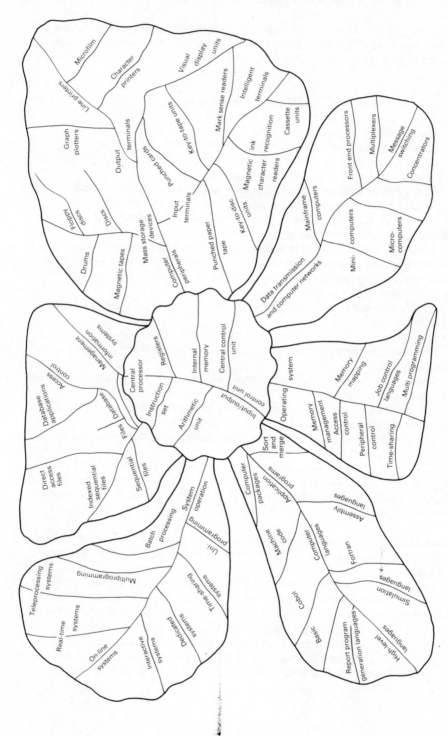

Fig. I.1 Graphic illustration of topics discussed in Part I

Part I
The System Hardware and Software

There is a proliferation of terms associated with computer technology. In this section an attempt is made to familiarize the reader, in descriptive fashion, with basic terms, concepts and tools of computer systems. Technical details are included, wherever necessary, for a better and complete understanding of the subject under consideration.

An operational computer system consists of the following essential elements:

1   Computer machinery, i.e. the basic computer hardware including associated peripherals used to process data.

2   Computer operating system, the interface between the computer hardware and the user, which controls and directs the use of computer resources.

3   Application programs which are run on the machine.

In order to make effective use of the computer, it is desirable that the users should have some knowledge of the computer hardware as well as the software. Hardware refers to the physical items which are combined together to form the basic computer configuration. The term software is used to describe the programs which control the activities on the computer, utility programs used, for example, to sort data or for creation and maintenance of data files, and application programs for individual tasks. This knowledge will make it possible for the users to appreciate the potential of computers in their applications, along with the constraints within which a computer system has to be designed and operated. The subject matter discussed in this section is graphically illustrated in Fig. I.1.

2
Computer Hardware

Introduction

Basically there are three different types of computers.
1   The analog computer in which basic data are manipulated in the form of continuous quantities.
2   The digital computer in which basic data are stored and manipulated in the form of discrete numbers.
3   The hybrid computer which combines the facilities available on the digital and analog computers.

In this book we will be concerned with the application of digital computers. Analog and hybrid computers are eminently suitable for certain specialized tasks, for example solution of differential equations.

Digital Computers

Essentially a digital computer is a device which can perform simple mathematical functions such as addition, subtraction, multiplication, division and comparison of data items, as well as a combination of these tasks. What distinguishes electronic digital computers from similar electro-mechanical devices is their ability to perform these basic functions at a very high speed. Some of the currently available digital computers can execute upwards of a million instructions per second.

For many years it was commonly believed that the first digital computer was developed in 1946 at the University of Pennsylvania's Moore school of Engineering. This particular computer named ENIAC (Electronic Numerical Integrator and Calculator), and invented by J. Eckert and W. Mauchly, consisted of 18 000 vacuum tubes, and did not use any moving parts to carry out calculations. ENIAC was capable of perfoming 5000 additions per second. However, it was revealed in 1976 that a British team led by Dr T. Flowers had built the COLOSSUS computer in 1943. This computer, comprising 1500 valves, was successfully used for deciphering coded German war messages.

The first generation of digital computers, based on the use of vacuum tube technology, had high memory access time. The memory itself was restricted. Also these computers were rather unreliable due to the limited lives of vacuum tubes; breakdowns were frequent, and very often it was impossible to complete a lengthy set of calculations. By the late fifties a second generation of computers, making use of transistor technology, was introduced. These computers were capable of executing thousands of instructions per second.

A third generation of computer systems, of much higher capacity and throughput has been available since the middle sixties. IBM 360 and 370 series computers are examples of third generation computer systems. Within each of these series there is a range of machines that are compatible yet have different powers, speeds and capabilities. These offer alternative input and output facilities as well as a number of storage media such as magnetic tapes, discs and drums.

With the second generation of computer systems, if a user decided to install a faster or bigger system of greater capacity and power to replace an existing machine it was necessary to rewrite all the application programs for the new computer. With the compatible series of computer systems it is possible to upgrade the system without the need to rewrite programs. Thus expensive new program development and lengthy delay in the implementation of the new system have been avoided.

The fourth generation of computer systems make use of solid state Metal Oxide Semiconductor (MOS) memory instead of core technology. Integrated circuit techniques are extensively used in present day computers. Transistors, diodes and resistors etc. required to perform the basic operations, can be fabricated together on a single piece of a semiconductor such as silicon. Integrated circuit devices are very reliable, and can be manufactured using highly automated machinery. As a result the costs of such units are low. Medium scale integration (MSI) and large scale integration (LSI) techniques make it possible to fabricate a large number of components on a single board.

The computer cannot process any data until an application program is prepared and stored in the memory of the system. Ordinary languages are not suitable for use with the computer system, and a number of special purpose languages have been designed for preparing computer programs. Within the computer memory the programmed instructions and data are stored in the form of binary digits (0's and 1's) which indicate one or other of the possible states of the computer memory.

The cost of basic computer hardware has been falling steadily as have the costs of peripheral units such as card readers, line printers, visual display units. However, the decline in the cost of peripherals has been at a much lower rate because the majority consist of electro-mechanical units.

The cost of preparing application programs has shown a steep increase during the past few years. Software costs are, in the main, linked to the salaries of professional programmers, and these salaries have been steadily rising. Software has to be tailored to meet the requirements of the users, whereas the computer hardware can be manufactured in large numbers using highly automated machinery.

Hardware

The basic building blocks of the digital computer hardware are:

1   Internal computer memory.
2   Arithmetic and logic unit.
3   Central control unit.
4   Input/output control unit.

These sub-units when connected together form what is commonly referred to as the Central Processing Unit (CPU) which is at the heart of all the activities of a digital computer. The processes which take place inside the CPU include transfer of information from the input unit to the internal memory, data manipulation, movement of information from the internal memory to the output unit, and shifting of information between the internal and secondary memories.

Central Processor

The term 'processor' refers to computer hardware units which interpret or execute instructions. The processors may be specialized or general purpose. An example of the former type is the 'input/output' processor which deals with input–output operations.

The central processing unit or central processor is a general purpose processor in which mathematical and logical operations are performed.

The CPU consists of a number of working registers, such as accumulators and index registers, and instruction registers in which instructions are executed. The processor has direct access to memory for data and instructions. The instructions are retrieved from the internal memory and located in the instruction register, when the interpretation or execution takes place. The address of the next instruction to be interpreted (or executed) is retrieved from the instruction address register, and the whole process is repeated.

The work of the CPU can be temporarily stopped by using 'interrupts'. This is a method used to draw the attention of the processor to the fact that an event has taken place. When interruption takes place, the status of the CPU is saved, and depending upon the type of intervention an appropriate command is executed. Interrupts are also used for carrying out functions such as the protection of memory and checking memory parity.

The CPU regulates the flow of data to and from the system, co-ordinates the activities of various sub-units, and executes individual instructions contained in the application program that has been previously stored in the computer memory. The instructions within the program represent the steps which must be taken to perform a given task, and no matter how sophisticated a computer might be it still has to be 'told' what to do.

The data are stored, in the form of binary numbers, in the internal memory, and the actual data to be manipulated are transferred, when required, from the internal memory to the logic unit which consists of one or more temporary data holding locations called 'registers'. Data manipulations are carried out in the arithmetic unit, which contains electronic circuitry to add and subtract etc. The central control unit issues the appropriate commands for the necessary manipulations by generating timing signals, which cause movement of data from the memory to the arithmetic unit, followed by data transformation and subsequent transfer back to the internal memory.

The speed of the CPU is usually measured in terms of the cycle time which varies from microseconds (10^{-6} seconds) for relatively slow computers to nanoseconds (10^{-9} seconds). Cycle time refers to the amount of time taken to carry out a given sequence of operations, for example, the time taken to retrieve data from the internal memory. In other cases the time taken to add two numbers of a given size and format is also used to define the speed of the CPU.

Fig. 2.1 shows a simple diagrammatic representation of a typical general purpose computer system.

Registers

The manipulation of data and execution of instructions within a digital computer, takes place in the registers. The data to be processed in the CPU is temporarily stored in them. The registers within a CPU have unique physical addresses, and the time required to access them is very low, typically between 2 and 5 nanoseconds.

The greater the number of registers in a machine, the more powerful it is. With more registers the number of instructions required to perform a task can be reduced and the programs are executed much faster. Some computers have general purpose registers, while in other computers registers are used as an extension of the accumulator for performing operations such as multiplication, division, memory access. A computer is more powerful if the registers are general purpose rather than restricted to specific tasks. Some of the registers are accessible while others are not. Therefore the power of the computer depends on the number of accessible registers and their flexibility.

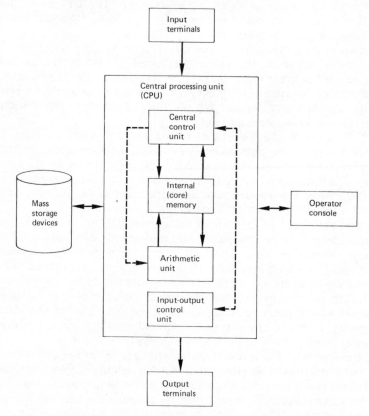

Fig. 2.1 Block diagram of the organization of a general purpose computer

Instruction Set

The capability and power of any computer is dependent on the range of instructions available. These may be of one of the three types as follows:

1 Instructions that move data between peripherals and various sections of the CPU.
2 Instructions which interrupt the sequential execution of a program and transfer the control of the program to alternative instructions according to whether or not a given condition is satisfied.
3 Instructions which operate on data and transform it according to the relationships specified in the program.

For typical scientific computers, used for 'number crunching' applications, the requirements for input–output capabilities are limited but the availability of instructions which simplify the manipulation of mathematical relationships results in a considerable increase in the speed of the computer. On the other hand a commercially orientated system requires only limited mathematical capabilities, with a much greater emphasis on an instruction set which can simplify the manipulation of large data files, and provide high input–output and data handling facilities. In most general purpose computers these facilities are combined and can be called up as required, under the control of the computer itself.

Some computers have large instruction sets used for performing a range of functions. For the same cycle time the bigger the instruction set the faster the rate at which programs are executed. As is to be expected, a computer with a bigger instruction set is more expensive than one with a restricted instruction set. Some computers have powerful instructions which are equivalent to a number of instructions on a smaller computer.

Complex instructions, such as multiplication, can be carried out either by using simple instructions (for example, repeated addition) or by means of special purpose circuitry within the computer. For instance, the floating point processor, whereby all types of decimal point arithmetic manipulations are carried out by hardware rather than software, may be used to dramatically increase the computational speed.

Internal Memory

Blocks of data for current use in the arithmetic unit, and program instructions to be used by the central control unit, along with any tables of required reference data and sections of the computer 'operating system', are all stored in the internal memory of the computer. Internal memory is a directly addressable and 'random access memory' (RAM).

This memory is also referred to as the core since it is often made out of ferrite rings. The ability to store information on ferrite core depends on the magnetic properties of the ferromagnetic materials. It is necessary to change from electrical representation of data to a magnetic representation before the data can be encoded on the magnetic storage element. Similarly for retrieving data, it is necessary to use techniques to change from a magnetic representation to an equivalent electrical representation. In recent years there has been the development of MOS (Metal Oxide Semiconductor) memory, which although faster than ferrite core is rather volatile. A power failure can result in the loss of the contents of the MOS memory. It is therefore desirable to have some non-volatile storage, particularly in the case of systems in which some software is stored for relatively long periods of time. Some computers include features used to protect the contents of the main memory as well as the registers, in the event of a power failure. It is thus possible to shut down the computer in a safe manner. Facilities also exist to restart automatically the computer on resumption of the power supply.

Alternative terms used to describe internal memory include main memory, memory, storage, main storage, and core. The memory is said to be erasable if its contents can be overwritten, i.e. the contents can differ from one occasion to another. Despite significant reductions in the price of component hardware, the internal or core memory is still a relatively expensive item, so that in most installations it is of limited size and the three categories of information referred to above reside only temporarily in the internal memory unit. External mass storage devices such as magnetic tapes, discs and drums, are used for storing large amounts of information only part of which is required for processing at any instant of time. Information is transferred as required between the internal memory and the external mass storage devices.

The internal computer memory is organized in the form of small units, each one of which can be uniquely defined by an address. Thus an internal memory location has two characteristics, namely, the memory address, which is the numeric code that describes its relative position in the internal computer memory, and the contents of the memory at that particular location.

Within a digital computer, the basic data unit is a 'bit' or binary digit which can take two states usually designated by 0 (zero) and 1 (one) respectively. Numerical data and alphabets are usually stored in a sequence of bits which are grouped together. The most commonly used field is a 'byte' which consists of an arrangement of 8 bits. Other arrangements consist of 12, 16, 24 or 32 bits collectively described as 'words'. The use of 'words' of different bit sizes often leads to confusion. The smallest unit of information which can be addressed directly depends on the design of the particular computer.

The amount of memory available on any computer device is referred to in terms of K words or K bytes, where $K = 2^{10} = 1024$ (In most other disciplines the symbol K refers to 1000). The computer storage units are usually organized so that they contain numbers of bytes or words which are an exact multiple of 2. Similarly a Megabyte does not refer to 1 million bytes. In fact it is equal to $2^{20} = 1\ 048\ 576$ bytes.

The internal memory of the computer can be extended, up to a certain limit, beyond the nominal memory by using memory add-on units. Such units are frequently used in modern computer installations.

Central Control Unit

The activities performed on the computer are under the direction of the central control unit, which in turn is under the command of the instructions contained in the computer program. The control unit acts as the switching section of the digital computer. Based on the programmed commands it decides when and how various operations should be performed. The control unit tells the arithmetic unit what to do, and informs it about the address of the location in the computer memory from which the necessary data should be retrieved. Once an instruction is executed, it is returned to the internal memory, followed by the retrieval of the next instruction from the memory and the whole process is repeated.

The central control unit also receives and identifies the signals which indicate any special conditions. It can also generate special signals. Clocks and other external sensors may be connected to the central control unit to provide special signals. Real-time clocks are employed in computer systems used for time critical applications such as process control, data acquistion, and also in 'multi-programming' and 'time-sharing' systems.

Arithmetic Unit

The arithmetic unit of the central processor carries out the actual work of performing mathematical operations (addition, multiplication etc.) as well as logical operations (branching and decision making, and the like). Logic circuitry made up from integrated circuits is used to perform these operations. Using the computer's ability to test, compare, and branch, programs required for the solution of complex problems and algorithms can be written. It is possible to make the sequence of instructions provide for several alternatives when there are changing values of one or more parameters. The circuitry can perform all the basic arithmetic functions, but with the cheaper computers only limited methods are used to carry out these functions. For example, circuitry used for adding numbers is also used to carry out subtraction, by using complements.

Input/Output Processor

Information resulting from the data manipulations carried out by the CPU must be communicated to the outside world. The input/output (I/O) terminals do not function automatically, but are under the control of the program. Some computers have independent processors which are dedicated to carrying out input/output operations. In other computers, mainly small ones, no such dedicated processors exist and it is necessary to disrupt the work of the central processor while such operations are being performed. Where specialized input/output processors are used, the CPU does not have to waste time on carrying out slow communications operations but instead can execute the programmed instructions at a very high speed. Input/output channels (processors) provide data communication between the peripherals and the internal memory. These channels are simple, slow, and less expensive than the CPU. 'Multiplexers' are used to connect a number of slow terminals such as card readers, paper tape readers, paper tape punch units, line printers to a single input/output processor.

The work of the I/O processor is usually started and stopped by the CPU which is in command. In effect the input/output unit acts as the interface between the computer hardware and the outside world.

Computer Peripherals

A general purpose computer system is not very useful without the availability of suitable peripherals which can be linked to the computer in order to input and

B

retrieve information, and to store data for short or long periods. A large number of peripherals is currently available. The advances in the field of data processing have reached a stage at which in most computer installations, the cost of peripherals attached to the computer often exceeds the cost of the computer itself. Computer peripherals, in the form of mass storage devices and input/output terminals, are used for communicating with the CPU and for enhancing the capability of the computer system.

Mass Storage Devices

In any computer installation, it is impractical to keep all the large files containing application programs plus current and historical data in the internal or main memory of the computer, because of the very heavy expenditure involved. It is, therefore, necessary to supplement the internal memory by means of mass data storage devices such as magnetic discs, drums, tapes, strips and cassette units. On some of these units the data are stored in a sequential format while on other units the data may be stored in a random manner. The information stored on these units can be brought into main memory, as and when required.

Magnetic Tapes

Magnetic tapes are typical of the sequential access data storage units. Historically magnetic tapes, plastic tapes coated with magnetic material, have been used for encoding information input to the computer, and also for storing the information processed and output from the computer. Tapes used for storing computer information are similar to tapes used in tape recorders except that they have seven or nine tracks on which data are recorded.

The information is stored on magnetic tapes in the form of binary digits or bits. Individual alphabets or numeric characters are represented by a row of patterns of bits across the tape. The data are read from or written onto the tape as it passes underneath the 'read/write' head. Some heads are in contact with the tape while others simply float over its surface. The second method is to be preferred since it results in a reduction in the wear of the magnetic tape, and increased reliability. On a nine track tape eight bits are used to represent a character, while the ninth is used as a 'parity bit'.

A 'parity bit' is an extra bit added, if necessary, to each row across the tape so that the total number of bits in the row is even for 'even parity' and odd for 'odd parity'. For example, if five bits are used to encode a particular character on the tape and 'even parity' technique is used, then a bit is added in the parity track so that the total number of bits is six, i.e. even. If, however, four bits are used to represent a character then for even parity there is no need to add a bit in the parity track. Parity technique is used to minimize the possibility of errors.

The information is stored in records consisting of a group of bytes rather than as single bytes. These records can be of arbitrary length and are identified by their physical position on the tape. Records are grouped together to form a block of information. A block of data is written or read from the tape in a single operation. Individual blocks of data are separated by an inter-block gap which consists of unrecorded tape between 0.50 and 0.75 inches long. The tape starts and stops at these gaps. Therefore the tape has to be allowed to reach nominal speed before any data transfer to or from the tape takes place. The overheads incurred in stopping/starting a tape can be minimized by increasing the maximum block length. Long blocks of data result in little space wasted in inter-block gaps, but also increase the core requirements for reading that block into memory.

The storage capacity of a magnetic tape is measured in terms of packing density, defined as the number of bits recorded on a single track of one inch of the tape. Packing densities typically vary from 200 to 6250 bits per inch (b.p.i.). The read/write speed varies from 4 inches per second to 200 inches per second. Thus in the case of a standard 2400 feet long tape, operating at a speed of 200 inches per second, it takes more than two minutes to traverse from one end of the tape to the other.

In order to retrieve a data item, stored on serially accessible magnetic tapes, it is necessary to search continuously through the whole of the magnetic tape until the required data item is located. On average it would be necessary to search through half of the magnetic tape. Magnetic tape access time has a large variance. In contrast the access time for internal memory is absolutely constant. Up to forty million characters can be stored on a standard tape of 2400 feet length and up to 300 000 characters per second can be transferred between the magnetic tape and internal computer memory.

To update information stored on a magnetic tape file it is necessary to create a new file on another magnetic tape. The information to be updated is read and stored in the main memory. The amendments are also stored in the main memory and applied to the records on the original file, and an updated file is created on the second magnetic tape unit. Fig. 2.2 shows this process of amending data files stored on magnetic tape.

Magnetic tapes can be overwritten and re-used. They are also relatively cheap and suitable for storing the large amounts of data required in the case of batch processed systems. However they are unsuitable for use with on-line systems due to the amount of time taken to locate the required data items.

Random Access Storage Units

The delay and limitations, inherent in sequential data storage and processing, may be overcome by storing data on random access storage devices such as discs and drums. The required data items can then be retrieved without searching through the whole of the unit. Almost all of the random access data storage units currently in use employ devices which rotate at high speeds. Information is stored in the form of tracks on the magnetic surface of the discs or drums and 'read/write' heads are used to read or write the data. The heads are located very close to the magnetic surface, but do not actually touch it. An addressing scheme is used to determine the location of any required data item and the head is instructed to move to that particular location and its contents are retrieved during the next revolution of the storage unit. A similar procedure is used for writing data on these random access devices. Since the data are stored on and retrieved from these units by means of an addressing scheme, and due to the random nature of these storage units, it is not necessary to store data in a predefined order.

The random access storage units can take a number of forms, the most

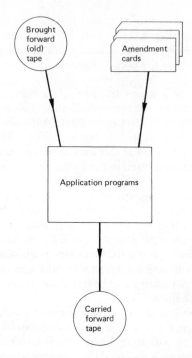

Fig. 2.2 Amendments of records on magnetic tape files

common being the moving head magnetic disc. The disc, which has a number of tracks on which information is stored, rotates continuously at a very high speed, and read/write heads placed over these tracks read or write information as required. The head moves in a radial direction, across the tracks, until the track containing the required information is located. The access time depends on the particular disc drive and is the sum of the time taken to locate the head over the appropriate track, and the elapsed time before the disc rotates to the correct position. Typical access time for moving head discs is of the order of 50 milliseconds. Often these discs can easily be removed from the drive. If simultaneous on-line access to all the data stored on random access devices is not required, it is possible to maintain large data files on separate discs which can be loaded, as the need arises.

The capacity of a disc pack varies according to the number of magnetic surfaces on the pack and is upwards of two million bytes. Fig. 2.3 shows a schematic representation of an exchangeable disc pack. The exchangeable disc pack is the most useful data storage medium, since it can be used with on-line as well as batch processed systems.

Sealed fixed discs which cannot be opened or removed from the drive, except for maintenance purposes, are also used. Due to the reduced danger of dust particles, fixed discs are more reliable than exchangeable discs.

In the case of fixed head discs, the read/write heads do not move, and individual heads are used for each track of the disc surface. The access time for fixed head discs is less than that for the moving head discs because the time taken

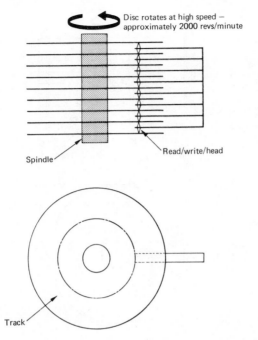

Disc rotates at high speed —
approximately 2000 revs/minute

Spindle

Read/write/head

Track

Fig. 2.3 Disc packs with one read/write head for each surface

for the mechanical movement of the head is eliminated altogether. Fixed head discs are also more reliable due to the elimination of a moving part. Such discs are sealed and cannot be loaded or unloaded as is the case with exchangeable disc packs. These discs have comparatively low storage capacity and are more expensive than moving head discs. Consequently the cost of storing data on fixed head discs is higher than for moving head discs. Fixed head units are generally used as swapping discs for moving programs to and from internal memory and disc when the computer is being used in a multi-programming or time-sharing mode.

Magnetic drums are also used as random access devices. The fixed magnetic drum consists of a large cylinder rotating about its axis at very high speed. The data are encoded on tracks on the surface of the drum. There are normally as many read/write heads as there are circular tracks on the cylinder, although some drums have more than one head for each track. Fig 2.4 shows a simplified representation of a magnetic drum. Blocks of information are transferred as required between the drum and the core memory, and it is possible to transfer a given block of information during the course of one complete revolution. In the case of drums with two heads per track, for instance, the time required to gain access to a given block of information is, at the very worst, only the time taken for half a revolution of the drum. Most magnetic drums have relatively small capacity and are expensive, but they do provide very fast access. In general, drums are used for storing frequently accessed programs, rather than large data files.

It is often necessary to use all three types of mass storage devices. The high

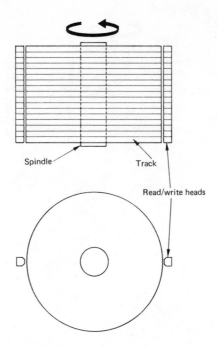

Fig. 2.4 Magnetic drum with 2 read/write heads per track

speed, but expensive, fixed head discs are used for storing programs, while moving head discs are used for storing large data files. It is almost always necessary to keep copies of various data files for the purpose of 'back up' storage, and relatively cheap magnetic tapes provide an economic way of storing this back up data.

Floppy Discs (Diskettes)

Floppy discs provide a cheap method of storing small volumes of data recorded on the oxide coated surface of a flexible mylar disc. The diskette is permanently packaged inside a protective envelope. Floppy discs are finding increasing application in the field of data preparation and entry. They can be unloaded from the drive, stored, and loaded later as required. The average access time for information recorded on diskettes is approximately 350 milliseconds and their typical storage capacity is 256 K bytes. In addition diskettes are more reliable than cassette units.

Input/Output Terminals

Input/output terminals can take a number of alternative forms, depending on whether or not it is desirable to keep permanent records that are easily understandable by the computer and/or user. Some terminals are suitable only for the separate functions of entering data into, or receiving information from the computer system, whereas other terminals can be used for both these functions. In addition, terminals can be distinguished according to their suitability for on-line or off-line modes of computer operation. A number of input/output terminals can be attached to the computer and used simultaneously.

Entering Information to the Computer System

Historically punched cards and paper tape have been used for entering information to the computer system. This process often takes a long time since it is necessary first to write information on coding sheets and then punch it on cards or paper tape. Often a second operator verifies the data on verification machines, in order to avoid data entry errors. Also the punch operators do not fully appreciate the significance of data and frequent punching errors are not uncommon.

Over the years a number of other methods of computer data entry have become available. It is now possible to enter data directly into the computer using special data entry or general purpose user terminals. Alternatively the data can be directly encoded on magnetic tapes or discs, by means of 'key-to-tape' and 'key-to-disc' units. The information can also be entered to the system during a normal dialogue between the user and the computer terminal. There are a number of advantages of entering data in an on-line mode of operation. For example:

1   The users, or dedicated terminal operators, are aware of the significance of the data, and the range of values which a particular parameter can take.
2   The data entry errors are immediately detected and appropriate corrections can be made.
3   The data can be captured at the source of its creation.
4   The need for verifying punched cards or paper tape, practised in a large number of data processing installations to ensure integrity of the data, is avoided.
5   The data files are always kept up to date. For a system of entering data by means of cards etc., the turn around time for coding, punching, verifying and then entering data to the system, can belong. It is often found that the punched data cards contain errors which are detected by input validation computer programs.

Disadvantages of on-line data entry include increased CPU costs, inefficient use of the computer hardware, and need for back up data entry equipment and standby procedures to be used in the event of computer breakdown.

Some terminals are specially designed to meet particular applications while others are more general purpose devices. The special feature may be a keyboard which reduces the need to press and enter frequently used characters. The characteristics of some of the frequently used peripheral units are now considered in detail.

Punched Cards

Punched cards represent the most widely used data input medium. Hollerith punched cards were in use long before the introduction of electronic data processing equipment. Most commonly used punched cards have eighty columns and twelve rows. One numeric, alphabetic or special character can be represented by one or more holes in one column of the card. The holes are usually rectangular, and are punched into the card, using a machine commonly referred to as a card punch, by depressing appropriate keys on a keyboard. The punch card keyboard is very similar to a typewriter keyboard except that all the alphabetic characters are upper case. Cards are interpreted by printing data above the columns in which they are punched. It is generally desirable to design a layout for the card. Blank spaces are often left in a card layout to allow for incorporation, at a later stage, of additional information. Colour schemes are also used to distinguish between cards used for different applications.

One card may be used to punch all or part of a record of data. Similarly it is possible to use all or some of the columns in a card. In the processing of transactions, the usual practice is to use one card to represent an individual transaction. Where this would result in a large amount of wasted space, a number of transactions are punched on one card. Similarly where the record consists of hundreds of characters, it is necessary to use a number of cards in order to enter all the data relating to that record.

Information encoded on punched cards is directly input to the computer using card readers, which operate at speeds of up to two thousand cards per minute.

A sorter can be used to sort punched cards into a sequence according to the information punched in a specified field.

The equipment used for punching, verifying and reading cards is relatively more expensive than that used for punched paper tape data preparation and reading.

Punched Paper Tape

Punched paper tape has been used for a long time for storing information. Now it is mainly used for capturing and entering data and programs to digital computers and other similar machines, for example Numerical Control (NC) machine tools. Information is stored on one inch wide paper in the form of a number of holes and the position of these holes determines the character. Most modern paper tape units use an eight track code. The holes in appropriate positions on seven tracks represent the alphanumeric characters and other special symbols. The hole in the remaining track is used as a parity bit. Photo-electric or slow mechanical readers are used to input data to the computer. The reading rate varies but is typically three hundred characters per second, although it can be as high as one thousand characters per second. The readers can operate in either 'start and stop' (asynchronous) mode, or the synchronous block mode.

The paper tape is not as wasteful, in terms of space, as a punched card. Records of variable length can be coded on punched paper tape, and it is not necessary to have leading blanks or zeroes. Only relevant data are punched and separators are used to indicate the end of a field. Another advantage of using paper tape as an input medium is that it can be produced in conjunction with other equipment such as accounting machines or teletypewriters etc.

Punched paper tape, as an input medium, suffers from a number of disadvantages. The paper tape is difficult to handle and is easily torn. It is not easy to edit. It is usually necessary to splice the tape, correct the error and then submit it for input to the computer system. Alternatively the errors can be detected and listed, and the correct data, in place of errors, can later be resubmitted for separate processing. Correction of cards is much easier; a new punched card can easily be placed in the correct position after a card containing errors has been detected and removed.

Preparation of data on punched cards and paper tapes is a slow, noisy and rather labour intensive process. The rate at which the data on cards or tapes can be input to the system is slow, and in 'uni-programming computer systems' the processor is often input/output bound.

Key-to-Tape and Key-to-Disc Units

Information can be directly coded on to magnetic media such as tapes or discs using key-to-tape or key-to-disc units. This method is much quieter than the punching of cards or tapes. In addition the tapes and discs can be reused thereby saving stationery costs. Also the rate of information input to the computer system is much faster than the rate for punched cards or paper tapes, leading to a reduction in the time during which the computer is input/output bound.

With key-to-tape units, the same equipment can be used for recording and verifying the data. A keyboard is used to enter data and the coded data are then verified by a different operator. Some key-to-tape units have additional facilities, for example the ability to locate particular records on the tape.

With key-to-disc units the information, entered by means of a keyboard, is directly encoded on the disc. On some key-to-disc units, information input takes place under the control of a small CPU which can be used to validate information before encoding it on the disc. Information recorded on disc is dumped, at suitable intervals, onto magnetic tapes. This is a security precaution to reduce the amount of time which might be required, in the event of accidental damage to the disc, to re-enter the lost information.

Mark Sense Readers

A number of devices currently available can be used to read characters of particular 'format' or to interpret marks made on documents. Other devices combine both these facilities. Optical character recognition (OCR) systems are used to convert data printed on documents into a digital form and then input the converted data to the computer. With such a system it is no longer necessary to carry out the labour intensive key punching associated with conventional data input techniques.

The OCR system comprises an optical scanner, a paper transport device and a recognition system. The variable information is recorded on the document to be input later, by means of 'mark sense' pencil marks. This method of recording the characters by hand is simple and information can easily be input to the computer. A soft lead pencil or a reprographic pen is used to make marks.

Documents may be rejected depending upon the quality of printing. A low reject rate is encountered with good quality printing, while in the case of documents printed using a high speed line printer the reject rate is high.

Magnetic Ink Character Recognition Readers

The magnetic ink character recognition (MICR) system is based on the use of stylized characters written using magnetic ink, which can be read manually as well as by the computer. Therefore information can be directly input to the computer. A number of MICR systems, which make use of American, British and Continental standards are available. In the American and British standards, stylized versions of numerics 0 to 9 along with some special characters are used. The Continental standard makes use of seven vertical bars, and the length of these vertical bars along with the interbar gap is used to represent different characters. Banks make considerable use of these systems. Documents associated with the MICR system can be read at high speeds using MICR readers.

Intelligent Terminals

Intelligent terminals, which represent a scaled down version of a minicomputer, are also used for data entry applications. The intelligent terminal contains the following sub-units:

(a) Central processing unit (CPU).
(b) A small amount of programmable memory.
(c) Buffer memory used for storing data before transmission to the host computer.

Data validation programs are stored in the programmable memory and input data validated at source, thereby reducing the load on the main computer system.

Intelligent terminals equipped with cassette tape or diskette backing store can be very useful in the event of breakdown of the main computer system. The cassette tape unit provides a fall back and the transactions data can be recorded on the cassette tape, and later transmitted to the main computer when it again becomes available.

Cassette Units

Cassette units are often used in place of paper tape and card readers. Cassette units of the type used for audio applications can also be used for recording digital data. These relatively inexpensive units are simple to load and data recorded on small cassettes are easily edited. Information is recorded either in the form of continuous or incremental records. However cassette drives are relatively slow and not very convenient for use with large data processing systems. The storage capacity is rather low, typically of the order of 300 K bytes. A major disadvantage of cassette recorders is the high error rate encountered during data recording and retrieval.

Visual Display Units

Visual display units (v.d.u.'s) provide one of the fastest means of interaction between the computer and users. A typical v.d.u. can display 24 lines of alphanumeric information on the screen. Each line contains 80 characters. Thus at any instant a full screen can show 1920 characters. The screen is refreshed, using local memory, at high rates. The majority of currently available v.d.u.'s use cathode ray tube (CRT) technology.

Most visual display units make use of cursors which indicate the position of the next input/output character. Software is used to control the position of the cursor which can be moved up, down, and sideways. A typewriter keyboard which has been modified to include some additional keys such as 'Delete', 'Erase', 'Copy', 'Backspace' is used to input information to the computer. Graphics and colour display facilities are also available on some of the v.d.u.'s.

The v.d.u.'s can operate in teletype (TTY) mode in which each character is transferred to the computer every time it is displayed on the screen. Alternatively the v.d.u.'s can operate in block mode whereby the complete message consisting of a block of data is transmitted to the computer. The time taken to transmit data to the computer is dependent on the data transmission rate. The TTY data transmission rate is 110 bits per second, while v.d.u.'s can operate at a much higher rate, typically 2400 bits per second. This high data transmission rate is particularly important when it is necessary to input or receive large amounts of data.

Some v.d.u.'s have all the control facilities built into them, and are of the 'stand alone' variety. Clustered terminals, consisting of a number of terminals used in a restricted area, can have one controller which may typically drive between eight and sixteen terminals. Some rather expensive display units with hard copy facilities are also available. Operational costs of such v.d.u.'s are high.

The availability of terminals such as v.d.u.'s makes it possible for a non-computer professional to be in direct and meaningful contact with a computer which may be located hundreds of miles away. The user can retrieve information, as required, and also assess the implications of alternative decisions made on the basis of information available to him. Thus the terminal, computer and user form a very powerful combination.

Visual display units with graphic facilities are particularly useful for production planning and control applications. Graphs which display, for example, sales forecasts, trends in inventory levels, changes in the level of product quality, work-in-process histograms, delinquency levels etc., can be comprehended easily by the management. A great deal of information can be conveyed quickly by graphs. It is also possible to examine the implications of changing the values of certain variables in a given equation, and display the results in the form of a graph.

Miscellaneous Input Terminals

Touch tone telephones consisting of a numeric keyboard are used for entering small amounts of variable data to the system. Additional keys facilitate the transmission of information to the computer. For example in the case of production orders, touch tone telephones are used to enter production order number, transaction code and the correct quantity. The data entered is immediately verified and in some cases an audio-response is generated. These voice response systems have a limited vocabulary, operate at slow speed and are rather expensive.

On the shop-floor special data collection terminals, from which completion of production operations etc. can be reported, are used in many advanced industries. A number of these special terminals can be controlled by a small process control computer. A plastic identification badge is used to identify the person responsible for entering the data. A punched card is then used to enter the fixed information, for example the current job being worked on, expected batch quantity etc. Variable data such as the quantity passed on to the next operation are entered via a keyboard. The keyboard may be just numeric or it may have all the alphabets as on a typewriter keyboard. Under program control the input data are validated and in the event of illogical data they are rejected; a person with much higher authority may be allowed to override this rejection of data by the computer. The transactions data are encoded on a suitable mass storage device.

Information can also be input to the system by a light pen used in conjunction with a visual display unit.

Computer Output

Computer print outs currently represent the most widely used method of presenting information, processed by the computer, for use by human beings. Computer generated output which will later be required for further data processing is generally output in a coded format on discs, magnetic tapes or paper tapes. In this section we are mainly concerned with output information which is visually inspected by the human user, and will examine various devices used for this purpose.

Line Printers

Line printers are widely used for printing out information which can be referred to at a later stage. However, since the early days of computing, improper use has been made of such devices. There has been a tendency, in the case of batch processed systems, to print out all the very large amount of data which might conceivably be required during the batch updating time interval. As a result the exception conditions, which have to be dealt with quickly, are submerged in a very large amount of detailed and often irrelevant information. Under these conditions it is highly desirable to list all the exception conditions separately.

Most line printers are characterized by printing rates within a range of 300 to 2000 lines per minute. The theoretical or nominal printing capacity is very rarely achieved in practice. The character set of most line printers comprises upper case alphabetic, and numeric characters. Some may also include lower case alphabetic as well as special characters. A typical print out line contains 120 to 136 characters. The majority of line printers have a printing density of ten characters per inch, and vertical spacing lies withing a range of six to eight lines per inch.

Impact printing techniques are used in some line printers. The printing is achieved by applying mechanical pressure to the character to be printed which presses the ribbon against paper. Additional copies, if required, can be produced using carbon paper or pressure sensitive paper. It is not necessary to use special stationery with impact printers, which are noisy and rather unreliable. In fact such printers often represent the least reliable component of a data processing system, and require frequent maintenance.

Non-impact printers make use of thermal or electrostatic printing techniques, require special paper and cannot produce multiple copies. In general they are quieter, faster and more reliable than impact printers.

Laser printers which can print at speeds in excess of 10 000 lines per minute are also available.

The price of printers varies according to the width of the line, the number of characters in the print set, and the speed of the printer.

Microfilm

Computer output on microfilm (COM) works on the principle of accepting digital data from the computer, converting it into analog form and then printing it on the microfilm at a very high rate. One COM unit can carry out the work of a number of impact printers. COM is generally produced in the form of sixteen millimetre rolls, cartridges and microfiche. Some models can be used for plotting graphs in addition to printing alphanumeric data. Where COM is used, it is no longer necessary to handle large volumes of paper which is the main feature of batch processed computer systems. It can also be used as an input medium, i.e. COM can be directly input to the computer. However COM is not reusable since it cannot be overwritten and provides a permanent record. Viewers are required to read information recorded on the microfilm. Where only one viewer is installed in a department, cross-referencing between two sets of microfilm data can cause problems which may be overcome by using expensive printer–viewers. Some of the more sophisticated microfilm based information systems have considerable logic facilities and make it possible to store and retrieve large amounts of information with ease.

Microfilm output is often used as an alternative to line printer output. This is particularly useful when it is necessary to refer to large volumes of information which is relatively static and does not require frequent updating. For example information relating to finished items in stock, sales forecasts, production schedules etc., updated daily, can be recorded on COM.

Graph Plotters

In the field of commercial data processing the use of graphic output has achieved only limited acceptance. It is a known fact that a large volume of printed output is not the most effective method of communicating information to busy managers. Graphs, histograms or simple trend curves are in general more effective than the printed word. The graphs can be displayed on the screen of visual display units equipped with graphic facilities. Graph plotters can be employed to produce hard copy graphs, using conventional pen on plotter facilities, or electrostatic techniques. With conventional plotters the plotting speed is dependent on the movement of the paper and not on the amount of information which has to be plotted. Characters or lines have to be individually plotted. The points resolution varies but can be as high as 1000 points per inch. Software is used to control the movement of the pen.

The electrostatic plotters provide more reliable operation than conventional pen plotters, especially when they are in continuous use. Addressable writing electrodes are used to plot the graphs. As the specially coated paper passes under the plotting head, the electrodes are instructed to store a charge on the paper. The charged paper is then processed using a liquid toner which contains carbon particles and this results in the appearance of black dots over charged areas.

Character Printers

The ubiquitous teletypewriter is an example of the character printers which print one character at a time. The teletype prints by striking individual characters against the ribbon which in turn is pressed against the paper. Although teletypes are relatively slow with a printing speed of ten characters per second, they do provide a relatively cheap method of printing out a small volume of information. Teletypes are directly linked to the computer, and the users can be in direct contact with the computer. In some installations teletypes are also used as console terminals and any error messages or abnormal conditions are printed out for later analysis. Mechanical paper tape punch and reader units are attached to some teletypewriters.

Other 'character by character' printers, much quieter than the teletypes, can operate at higher speeds of up to about forty characters per second. Such printers are now being used in preference to the ordinary teletypes.

Data Transmission and Communications Networks

A large variety of terminals, which access the computer using widely different methods and protocols, is currently available. It is necessary to transmit data over long distances when terminals are connected to a remote computer; telephone lines are often used to communicate this data. However most telephone lines can only transmit analog data whereas data used in a computer are in digital format. Modems are used to convert the digital data into suitable frequencies which can then be transmitted over the telephone lines. At the other end a similar modem is used to convert this data back into digital format which can then be processed by the computer. These modems carry out modulation and demodulation of the data and make it possible to transmit digital data over conventional communication lines.

The data communications take place using parallel transmission techniques whereby a number of bits are transmitted simultaneously. Alternatively one bit of data at a time can be communicated using serial transmission techniques. The latter technique is the usual method of data communications because most cables do not have the facilities to transmit a large number of bits simultaneously.

The data may be transmitted in one of three modes:

(a) Simplex mode in which the data are transmitted in one direction only.
(b) Half duplex mode whereby the data are transmitted in both directions, i.e. send or receive, but only in one direction at a given instant of time.
(c) Full duplex mode which makes it possible to transmit simultaneously in both directions. Full duplex facilities are only slightly more expensive than half duplex facilities.

Data communications networks are being increasingly used in modern computer systems. A typical computer network consists of one or a number of large computers, concentrators, front end processors, and a large number of simple terminals used for batch and interactive computing. The aim of the related communications network is to reduce data transmission costs and optimize the use of data communications lines. The communications networks also improve the system reliability and availability. In the event of the breakdown of a large computer, the front end processor, to which individual terminals are connected, can be linked to another computer, and service, although degraded, can continue.

Multiplexers

The transfer of data to and from the computer can take place in two ways:
1   Under program control and generally one word at a time.
2   Through a high speed channel.

Very often a number of slow speed terminals are attached to a central computer. This can be achieved by using separate data channels and individually connecting the terminals to the computer. However, in most instances the provision of separate data channels is inefficient since the transfer rate for typical low speed terminals such as teletypes is only 110 bits per second, which is very slow when compared to the high speed of the data channels. Even high speed peripherals operate at rates of only 9600 bits per second. The problem of inefficient use of data channels can be overcome by using a multiplexer or a multiplexing channel, which can handle a large number of slow peripheral devices. An asynchronous data communications multiplexer can control the transmission of data to and from a number of asynchronous terminals which operate at low speed. The multiplexer connects each of the terminals in turn to the computer, and the output from all the terminals is transmitted using the single channel. At the computer end a demultiplexer is required to reconstruct the data originating from individual terminals. Fig 2.5 shows a symbolic

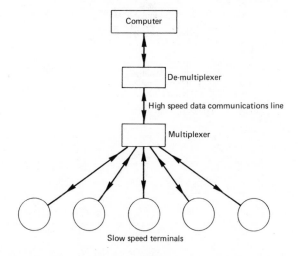

Fig. 2.5 Computer system with multiplexing facilities

representation of a system which makes use of a multiplexer and a de-multiplexer. The combined speed of the individual slow speed terminals attached to the multiplexer must be less than the speed of the high speed channel.

The use of multiplexer lines results in savings in the cost of computer hardware. The costs of the data communication lines required are also reduced.

Concentrators

Concentrators are often confused with multiplexers. While performing the same basic function as the multiplexer, the concentrators also optimize the performance of the data communication line. In the case of multiplexers, the high speed data line is divided into a number of slow speed data channels, each one of which is assigned to individual terminals. The concentrators make use of small computers and can store, in the buffer memory, the messages received from individual terminals. When the high speed data channel becomes free the messages can be transmitted to the main computer. Due to the availability of the data storage facility, the total transmission capacity of the individual slow speed terminals does not have to be less than the capacity of the high speed data line. The concentrators are also used for detecting and correcting any errors before passing the message to the computer.

Message Switching

The major difference between concentrators and message switching controllers lies in the fact that the concentrators collect data from a number of slow peripherals and transfer it to only one destination, whereas the message switching controllers collect data from several sources and then transfer it to one or several destinations. It is often necessary to store data received from one or

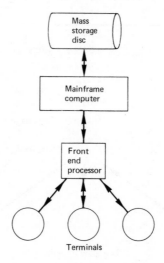

Fig. 2.6 A mainframe computer system with a front end processor

more computers and terminals. Consequently, secondary storage devices such as magnetic discs and tapes form part of a message switching system.

Front End Processor

Front end processors, which usually consist of small computers, are often connected by means of a standard interface to the input/output channel of a large computer system. The front end processor and related software are used for batching applications, reading in cards or paper tape, validating input data, handling peripheral devices, and communications processing. As a result the load on the main processor is reduced, and it can concentrate on the processing of application programs.

Programs for network and message control perform functions such as the polling and addressing of terminals. Similarly incoming messages are collected together in blocks, before the data in these blocks are transferred to the main computer. As a result the frequency at which the CPU of the main computer is disrupted can be reduced. Terminal interface programs are used to validate data input from terminals, by means of appropriate checks on the parity as well as format. Fig. 2.6 shows a mainframe computer system with a front end processor.

Microprogramming

Any programmed instructions performed using software can also be carried out by hardware, i.e. by wiring in the functions. Execution of instructions by hardware is very much faster but also more expensive. One instruction is used to start the execution of a large number of interrelated instructions. For example in the case of intelligent data entry terminals the microprogramming technique can be used for validating input data. If an input data item is unacceptable, then either it is automatically corrected according to some predefined instructions or the user can be informed about the errors in the input data.

When particular functions are performed very frequently it might be desirable to prepare the program using the hardware technique. This results in a highly reliable unit designed for that particular application. Microprogramming technique also makes it possible to modify the instruction set of the computer, and thereby customize the computer.

The microprogram is stored in the 'read only memory' (ROM) instead of being stored in the random access memory (RAM) like the remainder of the data. The program is wired into the computer's internal memory. The instructions can be read and used, but alterations are not allowed. It is therefore possible to avoid unintended modification to the program, although static electricity and high frequency transients, which might appear on the communication and power lines, can accidentally damage the program.

Programs wired into the read only memory can be changed only by altering the circuit boards which contain the necessary instructions. Rewiring of boards is a rather expensive process. Therefore the read only memory should be used only if no program alterations are anticipated.

The use of microprogramming and read only memory results in a special computer which is not software compatible with other computers.

Mainframe, Mini- and Micro-Computers

In technical terms there is a rather arbitrary dividing line between mainframe and mini-computers. Advances in computer technology mean that mini-computers of today are more powerful and faster than large computers of earlier years. It is, therefore, perhaps more meaningful to think in terms of applications for which particular computers can be used.

The single chip microprocessor, at the lower end of the market, is suitable for large volume applications in equipment such as photocopiers, telephones, computer terminals, data acquisition systems, motor car components. Micro-computers are suitable in applications with large memory requirements, as in commercial data processing or for industrial process control work calling for high speed and resolution.

Mini-computers can be used in an on-line mode, to carry out general purpose data processing applications. For example a mini-computer could be dedicated to the performance of production planning and control work in individual production lines. Such a system can provide up to date information for the production controllers and thereby help in the task of management decision making. Similar mini-computers can be used in other functional areas.

A large mainframe computer may be used to meet the data processing requirements in a number of different functional areas. The mainframe computer can also collect summary information, required for overall planning and production scheduling, from mini-computers used in individual production lines or departments. The selection of a computer configuration which can be used to meet the data processing requirements of any organization is a complex process and will be discussed further in later chapters.

3
Systems and Programming

Introduction

The effectiveness of a computer and the ease with which it can be utilized by potential users depends upon a number of factors such as the mode of operation of the computer system, the facilities available on the computer operating system, availability of high level language compilers, utility programs, computer packages, file handling software, and database management system software.

The above elements of computer systems are now examined in more detail.

Operation of Computer systems

The modes of operation of computer systems can be divided into three categories.

(1) Batch Processing

In batch processing the information about individual transactions, for example those relating to the work-in-process, is collected over a period of time, the data are coded, punched onto cards, paper tape, diskettes, magnetic tapes etc., verified, sorted into the order of records on the main datafile, matched with records on the main file, and then processed at fixed intervals of time, varying from days to months, to provide an updated output. Batch processing is an economic method for processing slow moving periodic information, and is particularly suitable for producing historical information to be used for analysis, long term planning and similar activities. Batch processing is also characterized by long turnaround times and large data preparation overheads. Batch processed systems cannot provide real-time answers at the place of origin of an enquiry, nor can they update information in real time. Large volumes of data used in batch processing systems can be stored on mass storage devices such as magnetic tapes and discs.

Remote job entry terminals, i.e. those not in the immediate vicinity of the computer, are sometimes used for remote batch processing. Following the completion of data processing on the computer, the computer output is transmitted back to the remote terminal. The availability of remote job entry terminals enables a number of users, at different locations, to have access to the computer. In all other respects remote batch processing is similar to local batch processing.

During the late fifties, when they were first implemented, batch processing systems had the attraction that their demands on computer time were low, computer time in those days being very expensive. Generally, however, batch systems are unsuitable for processing fast moving data.

(2) On-Line Systems

In on-line systems the transactions data are captured at source by means of special or general purpose data collection terminals, and the computer output, in a directly usable form, is transmitted to the area in which it is required. The users are in direct contact with the computer system by means of user terminals such as visual display units or character printers. In such systems the transactions may or may not immediately update the database or datafiles. The transactions may be stored on a separate file and batch processed later at some fixed time intervals. On-line systems in which the database is not immediately updated provide information about the state of a process, as it existed when the files were last updated. Typically the users can interact with the computer system to obtain information about such subjects as the status of work-in-process, availability of parts in stock, and the results of scientific computations.

(3) Real-Time Systems

Real-time systems have characteristics similar to on-line systems except that the datafiles are updated as soon as the transactions data are input to the system. These systems provide up to date information at all times, and consequently accurately portray the environment in which the system is operating. The system response time, i.e. the time interval between an input to the computer and the response generated by the computer system, varies from one application to another.

In the case of process control systems this response time might be a few milliseconds, while in applications involving simulation of a factory production line the response time might be a few minutes. The main distinction between a computer-based process control system and a general purpose system is that while the former can directly control a process, the latter type of system provides information which can be used to carry out planning and control. Most business systems are not as time critical as process control systems, and in general it is sufficient for the results to be returned within a few seconds. However, it must be borne in mind that relatively lengthy delay may lead to irritation and frustration on the part of staff awaiting information. In a typical manufacturing environment the computer, used to collect shop-floor data and update the database, might respond in a few seconds to enquiries about the status of work-in-process. Such a computer system is real-time in the sense that it provides up to date information at all times, and the decisions of the foreman or production controller are based on this knowledge of facts as they exist at that particular instant of time. It is unrealistic to describe a system as real-time if the information provided by it is not up to date, or if it requires more than a few minutes to respond to an input enquiry. An on-line real-time system should be capable of processing a large proportion of messages of an enquiry or file updating nature within a short time (say three to seven seconds). In ideal situations the response time of all data processing activities would be very short, but fast response can be very expensive to obtain. As a consequence it is often necessary to make compromises.

Running Programs on the Computer

Individual programs can be executed on the computer in a uni-programming, multi-programming, or a time-sharing mode. Let us consider these in more detail.

Uni-Programming

In a uni-programming computer system individual programs are run in a sequential mode and a program is executed until it is either complete, or errors are encountered in which case it is aborted. All the early computers were of this particular type. Even now some of the mini-computers can only operate in this mode. All the internal memory, apart from the area occupied by the 'operating system', is allocated to the job currently being executed. Small jobs may not occupy all the available memory and this leads to a wastage of internal memory. Also, a high proportion of business data processing requires the processing of individual transactions, and calls for the performance of a large number of input/output operations. Only a relatively small amount of time is spent in actual computations, since it takes a relatively long time (in terms of computer time scale) to input and output data using slow peripherals. As a consequence the CPU is idle for a large proportion of the total available time.

Uni-programming computer systems are suitable for some scientific data processing applications which require large amounts of actual computation with relatively small input/output requirements.

Multi-Programming

In a multi-programmed computer system, a number of jobs can be concurrently active but not necessarily be executed at a particular instant of time. Thus a job or program has been started but the processing has not been completed and the CPU is currently executing another job. The available internal memory is distributed amongst currently active programs. The main memory may be statically partitioned into a number of separate memory areas, and individual programs can be stored in these partitions by taking into account their memory requirements. Modern computer systems also make use of relocatable partitioning in which the memory requirements of individual programs are satisfied by dividing the memory in a dynamic mode.

In the case of computer systems with only restricted internal memory, other jobs may be held on the 'swapping' area of a disc or drum, and jobs swapped in and out of internal memory as required. Frequent swapping results in large system overheads since the CPU is idle during the swapping period. If it is desirable to run a number of programs concurrently the internal memory requirement is equivalent to the sum of the memory required by the largest program and the operating system. The program used to supervise the computer operations in a multi-programmed system has to include facilities for handling interrupts, a scheduling algorithm which can switch programs according to pre-defined rules, and tables in which information relating to the status of currently active programs is stored.

As an example let us assume that we wish to run programs for carrying out sales forecasting, work-in-process control, and order processing applications. Let us further assume that the main memory requirements for these programs are 20 K bytes, 40 K bytes and 36 K bytes respectively. If the memory requirements of the operating system are 48 K bytes, then there would be a minimum main memory requirement of 88 K bytes. In practice the memory requirements are upgraded to the nearest standard level of computer memory which in this particular case is 96 K bytes. If it is considered desirable to have all these programs resident simultaneously in the main memory, then the computer must have a main memory of 144 K bytes.

The processor initially assigned to the sales forecasting program will perform computations until it is necessary to carry out some input/output operations. At this stage the execution of the sales forecasting program is interrupted; while input/output operations are performed the current status of the sales forecasting program is saved. The CPU is assigned to the work-in-process control program, which carries out computations until it becomes input/output bound. The work-in-process control program is interrupted, its status saved, and the CPU assigned to the order processing program. When this program becomes input/output bound, the processor is assigned to one of the other programs.

Some of the scheduling algorithms used in multi-programmed systems incorporate features which, for example, can take account of the priorities associated with different programs, the computer resources required by the program, and the CPU time already used by the program. A multi-programming system can achieve a much higher processor utilization than would be possible in a uni-programming system. In some cases it might be possible to complete a number of jobs on a multi-programmed system in the time required to complete one job on a system without multi-programming facilities.

The processor time spent in waiting for input/output operations may be reduced by increasing the level or degree to which the system is multi-programmed i.e. the number of programs concurrently resident in memory.

Some computer systems operate in foreground–background mode. Background mode programs are computer bound, while foreground programs have priority and are input/output bound. In a multi-programmed system maximum utilization of available computer resources is achieved by transferring control from one foreground mode program to another, and when all the foreground mode programs are input/output bound the processor is assigned to the background mode program. This facility is particularly useful in 'interactive systems' used for processing large volumes of transactions.

Time-Sharing

The term 'time-sharing' or 'time-shared computer' may be applied to a system in which a computer is connected to a number of conversational terminals with which the central processing unit can communicate. Fig. 3.1 shows a simplified representation of such a system.

In a typical time-sharing system a number of users can use the computer simultaneously for updating files, enquiries, and producing large volumes of output on a line printer. If the data processing to be carried out in response to individual requests is trivial, then the response of the computer is very fast, and a user at a terminal has the illusion that he is the only person using the computer. This is because the computer can perform all these operations at a rate much higher than the rate at which most human users can respond. A scheduling algorithm is used to divide the CPU time amongst a number of jobs which might require processing at a given time. The time devoted to any particular job might be fixed or variable, and depends upon the scheduling algorithm. When the time slice allocated to a job is used up or the job becomes input/output bound, the processor is assigned to the next job waiting in the queue. The number of concurrent users is dependent on the total resources available on the computer. With a time-shared system used in a typical manufacturing company, the stock control clerk can enter data relating to items issued from the warehouse, the shop foreman can make enquiries about the status of work-in-process, while the engineers perform design calculations.

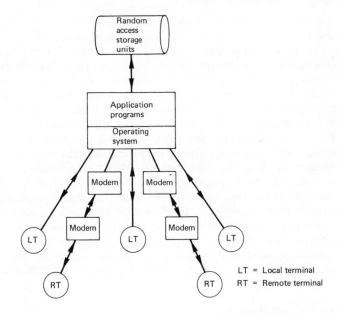

Fig. 3.1 A time-sharing system with local and remote terminals

The number of time-sharing systems installed is increasing all the time. Nevertheless, a high proportion of existing installations do not have such time-sharing or multi-access facilities. Other installations provide restricted facilities such as the creation and editing of files, and remote entry of jobs to be processed later. With multi-access time-sharing systems it is difficult to predict the computer load at any instant of time. However, in user terms such conversational time-sharing systems are more effective than batch systems.

Interactive Computer Systems

In interactive computer systems there is a high degree of interaction between the computer and users. Typically the computer will ask a pre-programmed question which is displayed on the user terminal, and then wait for the human response. The user may think for a while before entering a reply. The computer then carries out some processing, prints out more information, and then asks another question so that the entire process is repeated.

The extent to which a system is interactive depends on the rapidity with which exchanges occur and the amount of processing which must be carried out between exchanges. Interactive computing can only be carried out with on-line real-time systems. Moreover, in the case of a computer system with a small main memory, highly interactive work incurs heavy system overheads, since it may be necessary to swap different modules of an application program from the disc into the main memory for each interaction.

Dedicated Computer Systems

A dedicated computer system is devoted to a particular application and cannot be used for any other purpose. In the field of commercial data processing, airlines frequently use dedicated computer systems for seat reservation purposes. Airline agents at different locations are able to enquire about the availability of seats on particular flights and reserve seats. However, such a dedicated system cannot, for example, be used for performing any design calculations. In comparison, with a general purpose system it is possible to execute concurrently a number of different application programs using the multi-programming facilities.

There are advantages and disadvantages associated with both modes of operation. In the case of a general purpose system the efficiency of the computer is reduced while at the same time overheads are increased. The increase in overheads is due to the requirement for swapping different application programs in and out of the core memory. However, the overall utilization with such a computer system might be much higher than with a dedicated system. A dedicated system may be able to support a large number of active terminals simultaneously. Before a decision to use a dedicated system is made, considerable care must be taken to ensure that the special application under consideration provides sufficient load to ensure full utilization of the system.

Teleprocessing Systems

In teleprocessing systems computer power is available to a user by means of a suitable terminal, connected to a central processor, which may be located at any convenient place, so that work involving the use of a computer can be carried out as and when the need arises. Data communication facilities are used to connect the user terminal and the remote computer. In some cases it may be necessary to use modems or acoustic couplers, line concentrators and/or front end processors. The information is transmitted using public telephone network lines or private lines leased from the appropriate authorities.

The same computing power is available to the local and remote users. However, the remote user has to pay the additional cost of the communications line and the additional hardware and software which may be required. In some instances terminals are used for entering remote data which are batched together for later processing.

Computer Programs

A computer program consists of a number of sequential instructions which should be executed by the central processor in order to perform particular tasks in the required manner. Programs written by different people, to carry out a given task, differ appreciably in their content. This is because it is possible to take different approaches to perform a given function. Some programs occupy less storage space and can be easily read and understood, while other programs are often difficult to read and are inefficient because of large main memory requirements.

A good program is one which is simple, performs the required function efficiently using minimum code wherever possible and can be easily understood and maintained. Comment statements are often included as part of a program to make the program easily understandable, and for explaining any symbols or abbreviations used.

In big companies application programs often represent a large investment. As yet programs are still handwritten by skilled personnel although a substantial number of library programs is currently available. In the future it may be possible to use mass production techniques to prepare programs according to specified requirements. The characteristics of a program are defined by the size of the program, the program execution time, and the errors in the program. Almost all large or complex programs contain some errors. It is, therefore, important to ensure that all the programs are throughly tested and debugged.

Instructions are executed in a sequential mode as they are encountered. The only exception to this rule is when some branching instructions are included in the program code, in which case execution of the branched instruction takes place. The execution of an instruction sequence can also be stopped using the interrupt hardware signal.

The process of producing an error free program is iterative. Following the preparation of program specifications, flow charts which show the logic of the program graphically along with necessary data processing operations are prepared and checked to detect any logical errors. Discovery of errors at this stage reduces the time required to carry out the exhaustive program testing necessary to ensure accuracy and reliability of the program. It is often difficult to follow the logic of the program by reading through the program code. The standard symbols shown in Fig. 3.2 are frequently used for flow charting. The use of standard symbols means that anyone studying a flow chart can easily comprehend it. Large programs often result in complex flow charts.

Once the program flow chart has been verified, it can be coded in the selected programming language. The number of lines of code which the programmer has to write depends on the programming language used. In a low level assembly language it will be equal to the number of instructions which have to be executed by the computer. In a general purpose high level language the number of coded lines may be much less. After the program has been prepared it is input to the

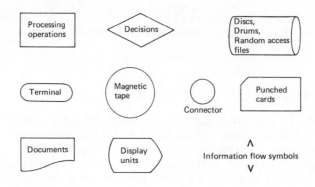

Fig. 3.2 Interpretation of standard flow charting symbols

computer and run with test data. At this stage some syntax errors may be discovered. Following the correction of syntax errors the results produced by the computer for test data are compared with the expected results. If the results do not agree then it is necessary to check the program logic and individual instructions for any errors, and the whole process is repeated until the program is free of errors.

The process of creating error free programs can be made less laborious and time consuming by testing them in an interactive conversational mode rather than the traditional batch mode.

Computer Languages

Computer programs can be prepared in a number of special languages which are precise and unambiguous, in contrast to natural languages such as English, which are full of ambiguities. The program can be written in;

(a) Machine code.
(b) Symbolic assembly language.
(c) High level problem orientated languages.
(d) Special purpose computer languages.

Let us look more closely at these languages.

Machine Code

At the lowest level the programs may be prepared in the machine language, i.e. the code which is executed by the computer. Machine language can either be written in the form of binary digits, i.e. as a sequence of 'O's' and '1's', or in hexadecimal notation which uses a base of sixteen. Within a machine code program it is necessary to keep track of memory locations. The codes used to initiate the performance of various instructions are dependent on the design of the central processing unit and the available instruction set, so that programs written in machine language for one computer cannot easily be used on a different computer. A good knowledge of computer hardware is a prerequisite for writing programs in machine code. Early computer users included specialist engineers and scientists who understood the intricacies of the internal workings of the computer and were able to prepare their programs in machine code.

The program written in machine code can be executed immediately. However, the preparation of machine code programs is an extremely time consuming process. Programs written in this form are prone to a large number of errors due to slips and oversights, and when lengthy programs are to be written the errors can cause severe problems. Debugging of machine code programs is laborious and some of the errors may not be detected easily. Also, these machine code programs are rather difficult to amend and maintain, particularly if they are lengthy. Documentation of machine code programs is burdensome since comment statements cannot be included.

Assembly Languages

The next level of language that can be used as an alternative is assembly language. Assembly languages use easily remembered mnemonics or symbolic representations of the machine instructions. Typical mnemonics are ADD, LOAD, SUB etc. The programmer has to write all the instructions which the computer must execute. Locations for the storage of instructions and data are allocated by the loader. Assembler programs are used to translate each mnemonic into machine language on a one to one basis. These programs are usually small and can be used easily on small inexpensive machines.

Assembly languages are closely related to the machine architecture and their structure resembles the structure of the appropriate machine language. Assembly language programs written for one computer cannot be transferred to another computer unless the instruction sets of the two computers are similar.

Some assemblers have error detecting facilities and also enable the programmer to write macros, with which one written line can correspond to several lines of machine code instructions. Compared to machine code programming, the preparation of programs in the symbolic assembly language is easier and less time consuming, and allows the programmer to have full control over individual instructions. In computer terms assembly languages are more efficient than high level languages. However the preparation, debugging and subsequent maintenance of assembly language programs is an arduous task. Operating system software is often written in assembly languages. The additional time required to run individual application programs, written in high level languages, may not be very important but the operating system software is used very frequently and its execution time is an important factor in determining the overall efficiency of the computer system. Typical examples of assembly languages are IBM ASSEMBLER, ICL PLAN, and DEC PAL11.

Cross-assemblers are sometimes used to assemble and debug an assembly language program on a large machine, and the resulting machine code binary program is executed on a small computer.

High Level Languages

Problem orientated high level languages can be used to solve problems in such fields as commercial, scientific and mathematical applications. High level programming languages simplify the task of imparting explicit instructions to the computer. Whereas some of these languages give instructions in simple English, which can be understood by a casual reader of the program, others are based on the use of mathematical relationships.

With high level languages a programmer need not focus his attention on the intricacies of the computer. He can thus concentrate on the way in which a particular problem is to be solved. These problem oriented languages are independent of the computer hardware, and it should be possible, at least in theory, to run these programs on other computers. Compilers or translation programs are used to transform the programmed instructions from the problem oriented language into the machine code which can be directly executed by the computer. Compilers are fairly complex and, depending upon the size of the program, the compilation process can take a considerable amount of time during which no actual computations are carried out. Also some of the compilers are large and require extensive memory capacity. Consequently some language compilers cannot be used on small computers. The programming language in which the program is written is known as the source language, and the instruction codes in the program are referred to as the source code. The language into which the program is translated, before execution, is usually known as the object language, and the resulting computer program called the object program.

Most currently available compilers have diagnostic facilities. The diagnostic programs detect errors in the code written by the programmer and then inform him about the type of error so that these errors can be corrected thereby leading to an increase in the productivity of the programmer. The availability of good diagnostic facilities means that the time taken to debug the program is reduced. Some compilers have additional facilities to detect minor errors and correct them automatically. Compilers only detect errors of syntax and not logical errors in the program; logical errors are detected only after the program has been tested using test data and the computer produced results compared with the expected results. Compilers are usually written onto the backing disc storage and called into main memory as and when required. Cross-compilers on high speed computers can be used to prepare object code for subsequent execution on small computers.

Many language compilers contain machine dependent features and thereby improve the efficiency of the particular programming language. However this optimization of the language efficiency can lead to other problems; the optimized language cannot be used on other computers with compilers which do not contain these machine dependent features.

Once a program is free of errors the object code can be stored on secondary memory devices. When it is necessary to execute these programs, the program is loaded following transfer from the secondary memory into the main memory.

Alternatively the computer can be instructed to examine the instructions contained in the problem oriented language and perform predefined actions. This mode of computer operations, in which the high level language statements are translated and executed immediately, is defined as the interpretive mode of program execution. This process is repeated every time it is necessary to execute the same instruction. Interpretive computing is a slow and rather inefficient process. Compiled instructions can be executed at a much higher rate than interpretive instructions.

Even simple statements in high level languages are complex from the computer hardware point of view, and one statement in a high level language is often translated into a number of machine code instructions. Programs written in high level languages require more main memory than those written in an assembly language to perform the same task. High level languages are inefficient and require long compilation time. However, there are a number of advantages in using high level languages. They are easy to learn and document. The programs are easy to understand and the programming and maintenance effort is reduced considerably. It is estimated that the use of the first assemblers reduced programming costs by a factor of five, and similarly the introduction of the first compilers reduced programming costs by another factor of five. The use of high level languages is increasing due to decreasing hardware and increasing software costs. Frequently used high level languages are described in the following sections.

Cobol

COBOL, an acronym of COmmon Business Oriented Language, is most frequently used for commercial computer programming. Its existence has greatly simplified the task of writing business programs. COBOL is intended to be a subset of English, and should be readable and understandable by most people. The file-structure facilities are powerful and can cater for information records of the types kept in most organisations–for example those relating to personnel, stock, production, and financial matters. This language is not suitable for large scale mathematical and scientific calculations. COBOL is easy to comprehend, learn and document, due to its simple English statements. However, it requires a large compiler and its main memory requirements are considerable. It is universally available and highly suitable for commercial batch processing applications.

Fortran

FORTRAN (FORmula TRANslation) is the most widely used language for scientific and mathematical computing. This language, originally developed by IBM, uses a mixture of mathematical and English statements. The programmer does not need to worry about individual locations and contents of the computer memory. Frequently used functions are either provided as a standard in the language, or are specially written in the form of sub-routines. Standard Fortran is not suitable for commercial applications, although business versions of Fortran with the necessary file handling facilities have been developed by some computer manufacturers and system houses. Fortran compilers are available for use on the vast majority of the currently available computers.

Attempts have been made to combine the features of COBOL and FORTRAN into single general purpose and comprehensive languages such as PL/1 (Programming Language One). PL/1 is a modular language and a user need master only the facilities required in his particular applications.

Basic

Beginners All purpose Symbolic Instruction Code (BASIC) is an interpretive language which requires minimal training. It is highly interactive. Improved versions such as Business BASIC, with considerable file handling facilities, are also available. Time-sharing BASIC supports a large number of terminals. Ever increasing use is being made of BASIC and its various extensions.

Report Program Generation Languages

Report program generator (RPG) is a much higher level language than Cobol. The required data processing is specified in the form of replies to a questionnaire. The files containing the required data, inputs to the program, calculations to be performed on the input data, and the format of the reports to be produced by the program are specified.

Data management systems are also used to perform standard data processing operations. Some data management systems generate code in a high level language such as COBOL and this code can be combined with some manually written code. The resulting code is used to perform tasks which cannot be carried out using the data management system.

Special Purpose Computer Languages

A number of computer languages have been developed for specialist applications. CORAL and RTL2 are used in real-time process/machine control, and communications applications. Languages such as APT, NELAPT, etc. are used in numerical control work. Due to the availability of these languages the amount of detailed work to be carried out by the numerical control machine programmer is reduced. Based on information provided to the computer, these languages automatically produce numerical control tapes. Cross checks are also performed to reduce the possibility of errors.

The standard software facilities provided by some computer manufacturers include a number of general purpose simulation languages which can be used to simulate a given set of conditions and test the implications of alternative strategies. Such languages may be used for simulating continuous as well as discrete systems. Continuous systems, which are usually associated with scientific and engineering applications, are modelled by a series of continuous equations, such as differential equations. In a discrete system, changes in the state of the system occur at discrete intervals of time. 'Monte Carlo' techniques are built into the simulation languages, and are used to incorporate random events and disturbances into the system. Popular simulation languages include GPSS, SIMSCRIPT, CSL, and DYNAMO, and they make the task of simulating an activity very much simpler than if the user had to write the whole program in a high level language such as FORTRAN. Also the availability of these languages helps the user in the construction of the simulation model.

For any application program which may be under consideration, the choice lies between programming in a low or high level language. Low level languages require less core storage and are efficient when the program is actually run. However, they are difficult to learn and are not portable. The trend is towards increasing use of high level languages. Within the high level languages, the programs may be written in a scientific high level language or a commercial high level language. The final decision is influenced principally by the task under consideration, the particular computer configuration, any standardization procedures operated by the organization concerned, and the efficiency of the language.

The efficiency of a programming language can often be measured in terms of the program development time, time taken to compile the program, execution time, time required to carry out necessary amendments to the program, as well as the time taken to document the program. Languages such as COBOL are easier to document than languages which make use of mathematical statements, and are therefore preferred. Compiled languages are often used in preference to interpretive languages. Other factors taken into account include the number of standard functions available in the language and the availability of interactive facilities.

None of the programming languages currently available is ideal for all applications, and new languages or improved versions of existing languages are continually being developed.

The economics of writing programs is also dependent on the form in which the program is to be used. If the same program is to be used on a large number of systems, and is not likely to require much updating then the assembly language is to be preferred on grounds of efficiency and low main memory requirements. If, however, the program might require frequent updating it should be written in a high level language. Most computer installations standardize on the use of one or more languages at high level as well as the assembler level. For example they might decide to use COBOL for commercial programming, and FORTRAN for scientific applications.

Overlaying

Main memory requirements of application programs often exceed the real memory capacity of the computer, and the program cannot fit into the available space. The programmer can overcome this limitation by using the overlay procedure. The program is divided into modules stored in separate overlays, which are read into memory as required and executed. Modules in different overlays cannot be simultaneously core resident. Program modules continuously required are permanently resident in the core memory. The overlay handler is used to define memory areas reserved for storing permanent parts of the program, and also for storing temporary sections. Memory segments used for temporary sections have to be large enough to contain the largest module to be brought into a particular main memory area. The programmer is responsible for designing his overlay system. Overlay handlers are included in a majority of modern operating systems.

Virtual Memory

Overlaying techniques allow only a limited extension of the main memory. Using virtual memory techniques an application program can be allowed to use a very large amount of memory. The maximum virtual memory available is dependent on the size of the directly addressable memory, which can run into millions of bytes, as well as on the total available backing store. Virtual memory is usually implemented using a combination of hardware and software techniques.

The file containing data is divided into a number of segments of variable length or pages of fixed length. The first segment or page is loaded into main memory and other pages are brought into main memory, as required, using disc to core swapping techniques. Page tables or segment directories are used to record the location of various sections of the data file, and to indicate whether a particular section of virtual memory is currently resident in main memory. Algorithms are used to convert virtual addresses into physical disc locations. Virtual storage means that loading and unloading of program sections and data is automatically and efficiently handled by the operating system. Large system overheads are incurred if the use of virtual storage facilities results in frequent disc to core swapping and vice versa. It is therefore desirable that application programs are written in a manner which results in minimum swapping. This can be achieved by referring to data items when the segment or page containing them is resident in main memory.

Utility Programs

Certain functions are common to all data processing installations; they include creation, editing, updating and maintenance of data files, system generation, and sort and merge operations. Most computer suppliers provide a number of programming tools as part of the system software package. On-line or batch text editors are used to replace, delete or insert individual characters or whole lines in a program. Frequently used mathematical functions and routines also form part of the standard software available on the system. When additional facilities or terminals are added to the system, the system generation programs are used to instruct the operating system about the new computer configuration. Software debugging tools are used to help the programmer prepare error free programs. Utility programs are also used for converting programs from one programming language to another, and for transferring information between various peripherals attached to the computer, for example for dumping data from one magnetic disc or tape to another, or for transferring information from disc or tape to the line printer. Sort and merge routines are two of the most frequently used tools in the vast majority of commercial data processing installations, and merit the following detailed description.

Sort Programs

In the processing of large volumes of data transactions which have been collected, over a period of time, in a random fashion, it is often necessary or desirable to sort these transactions into sequence, according to a specified key, in the same order as the file to be updated. Thus a data file is taken and processed to produce a new file containing the same records but sorted into a sequence. For example in order to produce an updated version of an inventory file it would be necessary to sort the inventory movement data in the same order as the old inventory file. It is estimated that in some computer installations, which use batch processing, as much as 30% of the computer time is used for sort operations. Most computer suppliers provide utility programs for sorting applications. A number of sort algorithms have been developed and the efficiency of a sort program, i.e. the time taken to sort a given set of records, is dependent on the algorithm used. Additional timing factors include the length of the sort key, and the size of main memory available for performing the sort. Sorting of data stored on discs is less time consuming than the sorting of data stored on tapes.

Merge Programs

Merge programs are frequently used in conjunction with sort programs for producing an output file from a number of input files. In a typical merge two files, generally consisting of the 'brought forward' (old) master file and sorted transactions data, are processed to produce a 'carried forward' (new) master file.

For example a new inventory file can be produced by merging an old inventory file and the sorted inventory movement data. The merging process consists of matching various input records, according to a specified key, and then manipulating the data to produce an updated file. The carried forward (new) file can be used as input to another program or it can be stored. Updated information is often printed out and used until a further update is carried out. The complexity of the merge programs depends upon factors such as the number of operations to be carried out as part of the updating process, the existence of input records on more than one file, and the technique used to identify the data on the master file. Fig. 3.3 shows an example of typical sort and merge operations.

The updating of a disc file is easy compared to the updating of a tape file. The records are retrieved according to the key number, the input data are matched and the file updated. New records are inserted using a separate routine.

Computer Packages and Report Generators

The writing, debugging, documentation and implementation of specially written application programs is an expensive, lengthy, and time consuming process. In many instances it is not necessary to prepare program code to carry out all the tasks which might be necessary in any computer installation. Computer manufacturers usually provide computer packages for performing commonly occurring tasks such as management of datafiles, housekeeping routines, and sorting of files.

Existing thoroughly tested computer packages are frequently used for carrying out mathematical and statistical computations, for example in numerical analysis work. The data processing task is usually fixed except for changes in the value of parameters. It is also possible to use computer packages for other routine applications, such as calculation of salaries/wages, or for the preparation of tax returns. However, even in this field, practices differ from one business organization to another, and many companies have written their own software. In general, computer packages are suitable for data processing applications in which procedures do not differ from one organization to another. The packages offer a cheap solution to the particular problems of a company considering the application of computers, since the cost of preparing the necessary software and documentation can be spread over a large number of customers.

Computer packages may have been written for use in a particular installation and modified later to make them more suitable for general use. Alternatively

c

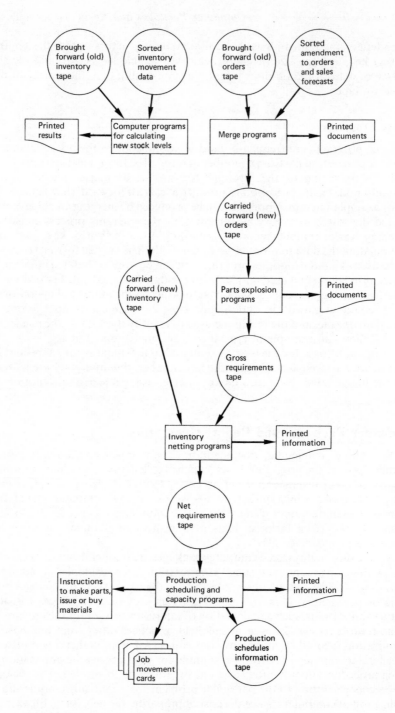

Fig 3.3 Typical example of the use of sort and merge routines in production planning and control applications

these packages may have been written, in the first place, for general use. A general purpose package comprises master files, with detailed records and some space for use by individual customers, along with application programs necessary to process the data files. The customer can either use the master files as specified by the software supplier, or he can create a new master file by including data in the specified empty positions. The supplied programs can be used directly to carry out the necessary data processing. Alternatively a new application program can be written, or amendments made to the supplied program, for processing additional data items. With some general purpose packages, options are provided to enable a variety of tasks to be carried out by specifying values of appropriate parameters.

For a number of other applications, production planning and control being a prime example, where procedures differ appreciably from one organization to another, it is often necessary to make large modifications to an existing package to satisfy fully the requirements of the users. Manufacturing companies, engaged in the production of a large variety of items, use widely differing production planning and control techniques. The file contents, i.e. the records and fields, along with the relationships and linkages between various data items also vary from one organization to another. Some computer suppliers and system houses offer a customization service, i.e. they carry out amendments to the package to satisfy the requirements of the users. Where it is necessary to carry out extensive modifications to an existing package, the time and money spent may be more than that required to create a new package. The management, systems analysts and programmers in the user organization have to spend a considerable period of time in reading various systems manuals, amending the supplied software, and deciding the procedures to be used for implementing the package. Other difficulties associated with the use of computer packages include inadequate system documentation, poor performance of the package, and errors in packages.

Software packages are useful for companies when they first start using computers for particular applications. The availability of packages enables them to start using the computer without have to wait for the long period usually required to develop the necessary application programs. It is often the case that after the companies have been using the computer for any appreciable length of time, they start developing their own software to cater fully for their own specific requirements.

A number of factors should be taken into account in selecting computer software packages. The obvious factors are the costs incurred in purchasing or renting, modifying, and running the package. The less obvious and frequently ignored factors are the package maintenance support received from the supplier, number of other users of the package, and system documentation standards. The use of a computer package might necessitate organizational and procedural changes so that users can fit in with the requirements of the package. Also the use of a package may not result in the realization of the full benefits of using the computer. It is therefore necessary to include the cost of benefits not realized in any discounted cash flow (DCF) analysis.

Report generators supplied by computer manufacturers etc. are frequently used in many data processing installations. The application of generalized report generation programs used in conjunction with a number of statements which define the processing necessary to produce the required output can lead to considerable benefits. For example in the case of work-in-process consisting of a large number of batches currently active on the shop-floor, it might be required to determine the location of batches, manufactured to a given specification number, which have not moved past a particular production operation. With the help of a general report generation program it is possible to prepare the required report using only a very small number of statements. In the absence of such report generation programs, it would be a mammoth task to prepare application programs for dealing with all the enquiries which might arise. Such general programs are necessary because in most instances it is rather difficult to predict the type of report or information which management might require at any given time; this is partly due to the fact that management is not always familiar with its own future requirements. With such a report generation system the user simply has to give information about the file on which the system has to operate, the selection of records according to predefined criteria, (for example the value of data in field 1 should be less than 40; value of data in field 4 should be less than 50 ... etc.), the data manipulations, and selected fields within the record which should be printed as part of the report. Thus data processing is carried out without the need to write full programs and the user need not even learn any programming language such as COBOL. From the user's point of view, these general report programs may be said to be very high level languages which he can easily learn. These programs can be of particular value in on-line systems; the users interact with the system, by means of conversational terminals, and specify their requirements.

The run time for such general purpose packages is greater than the time required to run a specially written application program. However, this increased computer run time has to be balanced against the cost and time required to prepare and test the special program.

Computer Operating System

During the pioneering days of data processing the computer executed one job at a time and was controlled to a large extent by the user. Individual users loaded the card reader, or mounted magnetic tapes, and generally operated the system at low speeds. The operation of the computer hardware and the very limited number of peripherals attached to the computer was controlled by a single unit. The speed of the computer was restricted to the speed of the slowest component in use. Thus, when the data was entered by means of a slow paper tape reader attached to a teletypewriter terminal, the computer was rather inefficiently used since it was input bound for a relatively long time.

The early computer applications were in the field of scientific and engineering computing. Most scientific applications, for example the solution of complex sets of differential equations, do not require a great deal of input and output data. Thus the majority of the CPU time is spent in actually performing calculations and only a small proportion of the total computer time is spent in input-output operations.

Small computer systems, particularly the ones for dedicated applications, are still used in this fashion. Procedures for handling individual peripherals are written as part of each program executed on that machine.

The computer hardware available today is extremely powerful and can be used for a very wide range of applications. Also modern computer systems are far too complex to be controlled by individual operators. People cannot be expected to make decisions and act at the fast rate dictated by the computer. Therefore, it is necessary to use systems and programs which can

(a) keep pace with the speed at which the computer works;
(b) handle the input and output to the system communication devices;
(c) schedule the individual tasks;
(d) allocate the space on the CPU to various tasks (application programs that are competing for processing at any given instant);
(e) memorize the state of individual programs, so that after the processing of a program has been interrupted, it can be resumed from the stage at which the interruption occurred.

Computer operating systems, comprising a set of programs and algorithms, are used to allocate and control automatically the facilities available on the computer. At any instant only part of the operating system is resident in main memory and the rest of the system software is stored on discs. The operating system, the software which acts as the interface between the computer and the users, lies at the heart of all the activities which take place in a modern computer system.

The operating system is an important constituent of the computer software, and the efficient running of a computer installation as well as the optimum utilization of the total resources is dependent on the facilities available on it. Most modern computers have associated operating systems. The trend towards making computer power available to professional programmers as well as non-professional users has been helped by the availability of sophisticated operating systems.

It is highly unusual for the users to write their own operating systems. Most computer manufacturers provide and continuously update the operating system software. The preparation of sophisticated operating systems requires many man-years of coding and testing.

Where operating systems, even the primitive batch ones, are used it is possible to control the peripherals automatically by writing all the required controlling and driving procedures. These programs can be stored in the form of sub-routines and called, as required, by programs which require the use of these peripherals, thus saving time and effort.

The operating system controls the computer resources, i.e. the processor, input/output peripherals, main memory, and mass storage devices, by keeping track of their status and allocating them to individual programs, according to a predefined set of rules. If necessary these resources can be reclaimed from one program and allocated to another.

Automatic accounting and control routines are included in most modern operating systems. Control routines are used to prevent unauthorized access to and use of the computer system. People wishing to use the system are requested to identify themselves by means of project number, user number, and passwords. If necessary passwords can be changed by the system manager. Depending upon the project number and user number, any particular user may not be allowed access to restricted system functions.

In many computer installations, particularly bureau services, users are charged for the computer resources used. The operating system keeps track of the computer time, main memory, disc or tape memory storage space, and input/output terminals used during the running of the programs, and users are charged accordingly. To prevent misuse of the system it is often desirable to cover the staff, stationery, maintenance and replacement costs of the computer installation by charging users for the service. Such costs may be real or notional.

Batch Operating Systems

A number of operating systems have been designed to run the computer in the batch processing mode. The main objective in the design of batch operating systems is to maximize the computer throughput. The program to be run is loaded and the language compiler used to translate the program into machine code. Any compilation errors are detected at this stage.

Most compilers print out error messages to indicate any syntax error encountered, and the current job is aborted. Otherwise the current job is run until the processing is completed when a new job is started. This particular sequence of operations refers to computer systems which do not have multi-programming facilities.

Multi-Programming Operating Systems

In a batch processed multi-programming environment the programs submitted for running are queued and a priority is allocated to them according to rules built into the scheduling algorithm. When processor time becomes available the program with highest priority is executed. A given program can be in one of three states at any instant of time:

(a) It is being executed.
(b) It is waiting for the completion of some external operations such as input or output.
(c) It is ready to be executed.

Thus a job submitted in batch environment may be in the process of execution when it requires information stored on data files. At this stage input/output operation is started and the program goes into an interrupt or wait state. On completion of data transfer a signal tells the operating system that input/output operation is complete and the program goes into a ready state.

A number of priority rules can be used to select the next program to be executed on the CPU. Some users may have a higher priority because they have not used the budget allowed to them. Alternatively the priority may be based on the time and resources required to run the program. In some systems the users are allowed to select the priority associated with their programs. In such cases they have to pay more for a higher priority and the better service.

The program to which the CPU is allocated can keep it until:

(a) It is complete.
(b) It is input/output bound.
(c) An error message is generated.
(d) A program with a higher priority wishes to use the CPU.

Alternatively a fixed slice of time may be allowed to competing application programs, in which case the CPU is allocated to another program after the expiry of the allowed time period. The CPU time required by different programs varies. Where equal time slices are allocated to competing programs, a program requiring a large amount of computation does not block the CPU to the disadvantage of other programs. From a functional point of view a multi-programmed computer system can be represented by Fig. 3.4.

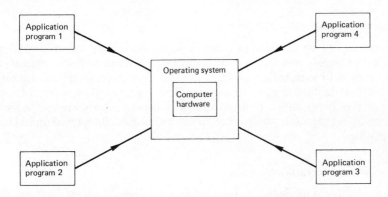

Fig. 3.4 Functional representation of a multi-programmed computer system

In multi-programming systems it is an essential requirement that the programs currently in main memory or secondary storage units do not interfere with the operating system software or other application programs. Interference between programs can lead to unpredictable results. The high degree of reliability required can be achieved using software although from the point of view of computer efficiency it is desirable to use hardware techniques.

Memory Management

Main memory is one of the relatively expensive components used in a computer system. Apart from costs the hardware design restricts the size of main memory which can be used in a particular computer system. A substantial portion of the computer main memory is permanently occupied by the operating system. The remainder of the memory is available for use by application programs. The number of jobs simultaneously resident in main memory is dependent on the maximum size of memory available, as well as memory requirements of individual programs. If unlimited main memory were available it would be possible to store all the programs in main memory.

In practice when a new program is input to the computer system it is temporarily stored on the backing store, and brought into main memory when it is ready to run. If the program is interrupted during execution, it is swapped back onto the disc. It is therefore necessary to keep track, at all times, of the available main memory locations and the programs which might be using it. Hardware as well as software memory management techniques are used to make decisions about the allocation of memory to competing programs.

'Roll in–roll out' techniques are used to save the status of a program on a reserved are of the backing store (Roll out), and allow the running of another program. The contents of the registers are also saved and it is, therefore, possible to bring back (Roll in) the program into main memory and resume the processing of the program at the stage at which it was interrupted to allow the other program to run. Frequent roll in–roll out results in high system overhead costs although these costs can be reduced using special hardware facilities.

Memory Mapping

Programs can be resident in contiguous memory areas or in a number of sections of the main memory. These memory sections can be in the form of pages of fixed length or variable length segments. The operating system software stores the program in available memory sections. The paging process has the advantage that a page of program can be fitted into any page of memory whereas in the case of segments it is necessary to carry out a search until a segment of required length is available.

Time-Sharing Operating Systems

The objective in time-sharing interactive systems is to provide fast response to a number of users linked to the computer. The response time increases with the number of jobs which are active at any instant of time. Too slow a response time breaks the concentration of the user and frequently leads to irritation.

In such systems one real processor is used to present the illusion that there are a number of independent virtual processors which are available to individual users. The number of such independent virtual processors is a function of the level of multi-programming available on the system. In the time-sharing environment the operating system functions at two distinct levels. At the lower level physical resources such as the CPU and peripherals are allocated, and the prime objective is to maximize computer throughput. At the higher system level the criterion to be satisfied is the service provided to individual users in terms of the system response time. Some of the resource allocation algorithms are simple while others use complex data analysis techniques and therefore take a relatively long time to allocate resources. At the lower level, algorithms are simple because the decisions have to be made very quickly, and a particular decision is effective only for a relatively short time. At the system level the objective is to satisfy a stricter and relatively longer time period service criterion, and more sophisticated algorithms are often used to provide the best possible service to a number of users.

In time-sharing mode the operating system does not always have enough information regarding the processing requirements of individual programs. Highly interactive jobs require high priority, while low priority can be allocated to less interactive jobs.

In some systems, program priorities can be altered as the processing continues. Operating systems which heuristically manage the control of resources and meet user requirements have also been developed. All the system activities are continuously monitored and the system is adapted, taking into account program requirements as well as the past behaviour of the users.

Control of Peripherals

In the majority of computer systems, peripherals such as line printers and discs can only be addressed through the operating system. Some peripherals can be shared while others can only be used with one program at a time. Examples of shared peripherals are disc, drum, and main memory. Dedicated peripherals include card readers, paper tape readers, and line printers. Once a dedicated peripheral has been assigned to a particular program it cannot be used by another program until released by the first program. When the processing of a particular program has been completed all the peripherals required by that program are automatically freed and become available for use by others.

The operational speed of peripherals is slow compared to the time taken to process the information contained in the single buffer associated with that peripheral; the resulting difficulties can be overcome by using additional buffers. In sophisticated systems the storage spaces on devices such as discs are frequently used for providing large buffer capacities. For example, data for use with application programs, input via slow card readers or paper tape readers, are temporarily stored on the disc. When time is available for executing the program, the data are read, as required, from the buffer.

Similarly the computer output is temporarily stored on the buffer space on the disc and printed when the relatively slow line printer becomes available. It is thus possible to use a number of virtual line printers although only one real line printer is available. Spooling programs are used to transfer information from discs to the appropriate peripheral. Line printer spooling is a feature of most modern operating systems.

The use of virtual terminals ensures that the CPU can be used at maximum speed, and peripherals are available at all times.

Job Control Languages

Most computer systems require a description of the programs to be executed. The job description includes parameters such as the program name, language compiler required, internal memory requirements, execution time, data files required to run the program, input/output peripherals required, priority to be allocated to the program, as well as instructions to show whether the program should be rerun in the event of a system breakdown. Default values are assumed for some of the parameters in the absence of proper specifications. In most modern computer systems physical devices, such as card readers, paper tape readers, line printers, are referred to by channel numbers, and the program is made device independent. The channel numbers associated with various devices form part of the job description. Operating systems include job control languages used for describing a job. The job control language informs the operating system about the resources required to run a job and allows the user to control the operation of his program.

Files

A collection of records used in a system is called a computer file. Files can be temporary or permanent. Temporary files are retained for only a short period of time, and are subsequently erased. The file retention time may simply be the time taken to run a program. As an example a program source code is translated into object code before the program can be executed. If it is not required to store the object code for subsequent use, it can be erased on completion of the processing. Temporary files are also known as work files.

Data Files

Commercial data processing mainly consists of processing large volumes of data stored as files. Individual data items, referred to as fields, are grouped in a logical record. The grouping together of fields depends upon a number of factors including:

(a) How often is it necessary to access the data?
(b) Keys used to identify data items, for example the information about quantity in stock can be retrieved by using the part number as a key.
(c) The relationships between keys, for example a part or component goes into a sub-assembly which in turn is part of an assembly.

In some cases it might be possible to use a single key to group together all the related data items. For example in an inventory control file, the part number can be used as a key and all the relevant data, such as economic batch quantity, quantity in stock, quantity on order, supplier's name, for a particular part number, can be easily retrieved. In an integrated inventory–production control system it might be necessary to retrieve data relating to an item not only by part number, but also using the assembly structure, and the suppliers from whom the particular item in question can be purchased.

Some record structures are of fixed length and fixed content. All the records in the file contain the same data items, and the same number of fields even though some of the fields may not contain any data. The only field which must have data is the one used as a key for referring to records held on the file. The problem of variable length field descriptions can be overcome using a field of the expected longest description; but at a later date it might be necessary to use an even larger field length. Blank spaces are left in the fields where data items are of short variable length.

Alternatively the file may consist of variable length records, and within individual records there is a variable number of fields of fixed length. The maximum record size is often predefined. This record structure is particularly useful, for example, for storing bill of materials data. A given part can be used in one of a number of assemblies. Similarly a component or material can be supplied by a number of competing suppliers. The record comprises a record key along with a variable number of fields. For each record in the file the system keeps a count of the number of variable fields attached to the record. This particular record structure makes good use of the mass storage devices since only the space required to store the data is used up. In some systems, the length of individual fields, within a record, is also variable in order to cater for the different lengths of the names and addresses of the employees of a company, for example.

Very often when data are stored as files an overflow occurs, i.e. the storage space available is inadequate to cater for the size of particular records. During the file design process, provision should be made for dealing with such eventualities. Usually the key field is duplicated in another record and the variable data are stored in the remainder of the record.

A directory is used to record information relating to the status of existing files. Typically such information comprises the file name, file generation number, the size of the file, location of the file in the case of random access devices, owner identification, and any information relating to privileged access. If a password is associated with a file then any user trying to open the file must give that password. Directories themselves are stored as files on the storage medium.

File Organization

A number of datafile organization techniques are currently used.

(1) Sequential/Serial Files

Magnetic tape files, and some disc files are organized in this fashion. The records are physically stored in the same order as the key number, and it is not necessary to use any index or algorithms to locate a particular record.

The usual method of updating such files is by copying the whole file from one device to another. The transactions data are sorted in the same sequence as the master file, the records are matched and subsequently merged to create an updated version. Old copies are retained for use in the event of file corruption or system breakdown. Sequential files are only suitable for batch processing and cannot be used with on-line systems. This file organization technique also incurs system overheads in terms of the unwanted information which has to be copied from one device to another. Its main advantage lies in the fact that such files can be stored on relatively cheap magnetic tapes.

(2) Indexed Sequential Files

Indexed sequential file organization is frequently used for batch as well as on-line systems. The records are arranged in the same sequence as the key number, and an index or table is used to define the actual location of these records on the random access unit. In the case of systems with a number of disc units, a double index is maintained. The first index shows the disc pack on which the file is stored, and the second keeps a record of the track on which the required data are located.

These files can be processed sequentially as well as by referring to the individual keys. The amendment of individual records in indexed sequential files is much easier than amending records in sequential files. It is only necessary to define the key number of the relevant record, and no reference is made to other records which are not required. The records are amended in place, and the new data overwrites the existing data.

File security is achieved by dumping the whole file onto another disc or tape. The amendments and updates can be written onto another storage device, and in the event of a system breakdown the datafile is reconstructed using a previous version of the file and the updates. The major problem occurs when it is necessary to insert new records. Overflow areas can be used to store additional records temporarily, and the files are periodically reorganized. These files cannot be stored on magnetic tapes.

(3) **Random (Direct Access) Files**

In random or direct access files, stored on discs, algorithms are used to define the relationship between the record key and the physical location of the record. Individual records are retrieved without the need to search through the file. There is no direct relationship between contiguous records. Additional records are simply added to the end of the file which is updated by overwriting existing records.

The time required to determine the physical location of data, in direct access files, is less than the equivalent time required for indexed sequential files. This is because it is not necessary to search through the index or tables.

Individual records cannot be located without a knowledge of the key number. Similarly the key of the next record on the storage medium cannot be determined. If it is necessary to print all or some of the records in a direct access file, then the file has to be sorted according to the specified key number. The record access time, dependent on the disc arm movement, is reduced by first determining the physical locations of records to be accessed and minimizing the arm movement.

Direct access files use the available storage space efficiently and are suitable for on-line systems in which the record access time is an important factor.

File Processing

Processing of computer files typically involves operations such as file creation, insertion and deletion of records, amendment of existing records, and file enquiries.

(1) *File Creation*

Files are initially created by collecting together all the relevant records. For example to create an inventory file for all currently active parts it is necessary to gather data such as part number, part description, quantity in stock, quantity backordered, economic batch quantity, bin number, suppliers. Special programs which include data validation facilities are sometimes used for creating master files.

(2) *Insertion and Deletion of Records*

It is often necessary to insert new records in an existing file. For example as new parts are added to the list of products manufactured and/or stocked by a company, this has to be reflected in the inventory file. Similarly it is desirable to delete records from files in order to minimize the file size, thereby saving storage space. For example, inventory records relating to obsolete parts should be deleted unless this information is required for historical analysis.

(3) *Amendment of Records.*

Updating of records, to reflect changed circumstances, is the most frequently encountered file processing operation. For example as an item is issued from stock, the inventory file has to be updated by subtracting the quantity issued from the quantity previously held in stock.

(4) *File Enquiries.*
Management requires up to date information for effective decision making. On-line systems include facilities to access information without amending the file in any way. For example the sales personnel may wish to enquire about the available free stock of an item before quoting a delivery date to customers.

Operating systems also include facilities for storing large data volumes in the form of files. The data are stored in blocks which are chained together. Magnetic tapes are the most widely used medium for large scale data file storage. More than one file can be stored on a single tape. The title block contains the name of the file, date on which file was created, the file generation number, and the number of blocks on the file.

Frequently used data are often stored as disc files, and a disc directory is used to record the location of data items. In the absence of a disc directory it would be necessary to scan the whole of the disc to retrieve required data. In multi-programming and time-sharing environments, files belonging to many users are stored on a single disc. When access to a particular file is required the information recorded in the directory is used with a search algorithm to locate the block containing the required data. The file, referred to by its name, is opened and depending upon the user's access privileges, he may be allowed to read, write or delete the data. In a time-sharing environment the type of access allowed to individual users can take a number of forms as follows:

1   The file may be simultaneously opened for writing and reading facilities accorded to any number of programs. However, if a block is being written on then access to other users is temporarily denied.
2   Alternatively any number of users can read the file but only one user has writing privileges.
3   In other systems a file can be open for reading or writing by any number of programs. However the file cannot be opened for simultaneous performance of both these tasks.
4   It might be necessary to further restrict access to the file and open it for reading or writing by only one program at a time.

Where a number of programs need to retrieve data stored on a disc with a single movable read/write head, the operating system should attempt to minimize the number of head movements.

Database Systems

Most manufacturing companies, in common with other types of organizations, store, manipulate and use large volumes of data which cannot easily be handled by conventional manual methods. The data have to be stored in a manner that provides relatively easy access. All the data used by a company can be kept in two different ways–as a small number of large files, or a large number of small files. There are advantages and disadvantages associated with both these concepts. Where a small number of files are used to store data, the mass storage devices can be used efficiently, because it is not necessary to repeat the data and related keys in different files. In addition, the task of maintaining, correcting and updating the files becomes easier. On the other hand, with the alternative method of storing data in the form of a large number of small files, the individual files can be 'tailored' to meet the specific requirements of different applications, so that the sorting of files for different applications is reduced to a minimum.

Conventional methods of storing data in the form of individual files used in conjunction with one or more application programs, suffer from a number of disadvantages as follows:

1   The file system is inefficient because it leads to a duplication of data. Let us consider, for example, the information relating to a manufactured sub-assembly made up from a number of components. This information is kept in a bill of materials file. The sub-assembly may be used in a number of assemblies, and this information would be kept on a file which shows the assemblies in which individual items are used. Design specifications will be stored in an engineering data file. Similarly, information relating to quantities currently in stock, on order, economic batch quantity, etc. is kept in the inventory file. The routing file lists the manufacturing operations. The 'assembly build program' file shows the quantities required over the next six months, for example. Other files will be required to store product costing and spares data. Thus the information relating to the sub-assembly is repeated in a large number of files, each one of which has been designed to store a particular data record.

2   If all the files which contain a particular piece of information are not simultaneously updated due to a lack of discipline or oversight, then the problem of data inconsistency occurs. For example increased material costs result in an increase in manufacturing costs and this product cost information is required in the inventory file for updating economic batch quantity data, and also in a sales order processing system used for preparing customer invoices. Unless discipline is rigidly enforced the task of controlling the data becomes difficult.

3   When new computer applications are envisaged then it might be necessary to create yet another file in which some of the currently available data are again duplicated.

In recent years there has been an increasing trend towards the use of 'database' systems. A number of definitions of a database have been proposed. The essential requirements of a database system are that the data should be organized and integrated so that they represent the natural relationships between various data items, and can easily be accessed and used. Also the database should minimize the duplications of data. Once such a database has been created, it can be used for a wide range of applications, some of which may not have been conceived at the time of creation of the database. Software is used to manipulate the data contained in the database, and this is known as a database management system (DBMS) software. Special high level languages, which can be used to simplify access to this data, have also been designed.

In database terminology a data item or element is the smallest unit of data in a database. A record is a collection of data items. The records identify and provide a description of a particular item, and are used as input–output units in individual application programs. A group, comprising a number of elements, is a sub-set of a record. A set defines a relationship between a number of records and makes it possible to obtain acces to related records.

A data item can have relationships with one or a number of other data items or records. These relationships are represented by means of physical adjacency or pointers. Relationships or pointers may be undirectional or bidirectional.

In most database systems an attempt is made to ensure that the data are not duplicated. However complete elimination of data redundancy may not be possible or even advisable in some cases. For example, for providing adequate response to enquiries, and for increasing the computer throughput rate it might be desirable to include the same data item twice.

All the authorized users who wish to use a particular data item can access it, and operate on the same data values. Different departments have responsibility for updating various data items. The responsible departments input values to the various data items, and these values can later be retrieved by other user departments. The values stored in a database are abstract until the users interpret them by means of suitable relationships.

A data description language defines the basic functions necessary in the management of a database, namely the form in which the data are stored on the database, the relationships between records, and the way in which the data can be retrieved.

With database systems the task of writing application programs becomes much easier. Application programs which do not make use of database techniques are complex since the data required in such a program may have to be retrieved from a number of data files. In the event of any changes in the file structure the application programs have to be amended. With database systems the application programs are independent of the database structure, the task of data retrieval being performed by the database handler. The same data can be retrieved in a number of different ways.

The access to the database can be by means of interpretive statements which are immediately executed. This is useful for on-line enquiries but not for large systems which make use of batch processing techniques. In other systems simple statements, which are first translated into a high level language and then compiled, are used to obtain access to the database.

A database handler or database manager is used as the interface between the operating system and the user programs. In turn the operating system acts as the interface between the database manager and the database. Thus when it is necessary to retrieve some data, the request is passed from the application program to the database handler which determines the physical location of data, and the operating system is instructed to retrieve the data. The retrieved data are written on the system buffers, and used as required. The necessary data manipulations can be carried out using a special data manipulation language or a conventional programming language such as COBOL, FORTRAN, or PL/1. Some database management systems are self contained, while others permit the use of host programming languages. The main advantage of a self contained database is that the language is specifically designed for that particular application. However these self contained systems may not be as comprehensive as the systems with host language facilities.

Database Access Control

The formulation of a database is an arduous and time consuming process which calls for standardization of all the data items and a clear understanding of the relationships between individual data items. It is therefore necessary to implement a discipline which ensures that standard codes and procedures are used. A database, once created, contains a large volume of information which is retrieved, manipulated and used by a number of people in different departments.

It is imperative to ensure that only authorized personnel are allowed to access all or parts of the data. Database access controls can be enforced, for example, by checking that the particular user is authorized to use a particular terminal. He can be asked to identify himself by means of identification cards or passwords. Access to records or data items can be restricted to individual users or particular programs. Also the access privileges allowed to different users can be of different types; some users can be allowed to read, manipulate and delete the data, while others may simply be restricted to reading the data. In some organizations a database administrator controls the access to the database and also ensures its integrity.

Most modern database systems have very comprehensive facilities not just for the organization and ease of access to the data, but also for preventing unauthorized access. Database reorganization facilities are used to periodically reorganize the database, and make efficient use of the available mass storage space. Facilities also exist to carry out a system recovery in the event of a system breakdown. Rollback procedures, included in some database systems, can be used to cancel the recent updates to the database, and return it to the state in which it existed before these updates were carried out.

Database Applications

Database systems are particularly useful for manufacturing companies because of the complex relationships which exist between the sales forecasts, orders received, inventory held, requirements planning, capacity planning, loading and scheduling of work, the work-in-process, and product costs. An integrated manufacturing database has considerable potential in all these functional areas.

The database approach is most beneficial when used in an on-line mode of computer operations. Batch systems do not make proper utilization of all the facilities available in a database management software. However, for maximizing computer throughput it might be necessary to use the computer as well as the database in the batch mode. In such circumstances the best solution is to use the computer and database in on-line mode for some applications, in particular the ones involving frequent updates of data, and batch mode for other applications.

Management Information Systems

During the early years of commercial data processing the emphasis was simply on the processing of large volumes of data. Typical computer applications were pay-roll and accountancy calculations. The sucessful use of the computer for these applications resulted in its use for providing relevant information to management, and for controlling events. This led to the development of what is commonly referred to as a management information system (MIS). Management information systems are generally intended to provide timely reports used by senior management for decision making. The main requirements of a well designed management information system are:

(a) A comprehensive database maintained on random access units.
(b) Access to the computer and database by means of local or remote conversational time-sharing terminals.
(c) A quick response to enquiries made by users.
(d) A flexible system which can be easily operated, using simple statements, without any requirement for specialist programming skills.

Although the use of management information systems has grown out of requirements of senior management, such systems make an even bigger impact on line management at operational levels. For example, the production management personnel require up to date information about the day-to-day operations for frequent decision making. Technological improvements along with cost reductions have made it possible to apply computers to these application areas. Any management information system must be dependent upon the needs of the users rather than the state of the technology, although the technology available decides the constraints within which the system must be designed.

Information requirements often change over a long period of time. Also most managers work without explicitly understanding the information which they use in arriving at decisions. Therefore the managers are not always able to define their information requirements. In the case of batch processed systems it is particularly important that in the absence of specified information requirements, all the possible information is not printed out. This has, in the past, been the cause of the failure of a number of otherwise well conceived systems. Most managers are busy people who do not have the time to search through a large volume of printed information and often go back to using informal systems.

Currently available technology makes it possible to design flexible management information systems. This flexibility is highly desirable so that the MIS can be used by a number of different people whose requirements differ according to their background and positions. The flexibility of a MIS also ensures that when a new manager takes over an existing job, it is not necessary to produce a new system.

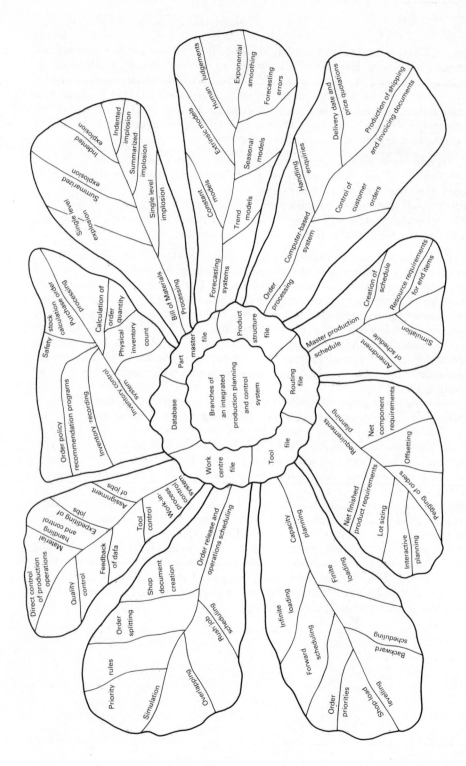

Fig. II.1 Graphic illustration of main subjects discussed in Part II

Part II
Production System Modules

An integrated computer-based production planning and control system consists of a number of application modules which are separately developed following a definition of the overall role of the computer in the production function, and the specification of the requirements for individual modules. The sequence in which various activities are implemented on the computer will vary from one organization to another according to their parochial requirements, the need for an improved system in a particular functional area, and the financial return on the capital and manpower resources invested. The important features of the application modules, illustrated in Fig. II.1, are now considered in detail in the following ten chapters.

Part II
Production System Models

4
Production Database

Introduction

In the manufacturing environment it is usually necessary to maintain a very large volume of data which is accessible to and used by a number of company departments. Even in a small manufacturing organization there might be as many as 30 000 individual items for which data, such as the unique part number which identifies the product, part description, quantity on hand, manufacturing lead time, reorder point has to be maintained. Similar data are maintained for raw material items handled by the company concerned. It is also necessary to keep a detailed record of the available manufacturing facilities. Most companies divide their manufacturing facilities into groups of machines which can carry out largely similar tasks and these groups are usually referred to as machine or work centres. For production planning applications it is necessary to maintain records of total capacity available and capacity which has been allocated to the work-in-process as well as production orders waiting to be released to the shop-floor.

For manufactured items it is also necessary to keep data relating to a sequence of production operations, carried out to produce given items, along with a variety of data associated with the particular manufacturing process. This data would include, for example, the standard set up time, run time, transit time, description of the operation, the machine or work centre at which the operation is to be carried out, the jigs and tools used during manufacturing and how the operation is to be performed.

Products of any complexity are usually assembled from a number of components. A particular component may be used in more than one end item. Consequently details of relationships between products and components have also to be recorded.

This data can be organized in the form of a number of simple files or preferably as a number of linked files, using the direct access capability of the discs, so that the data can be retrieved in the required format. In the past the usual practice was to store such data, particularly the large files, on magnetic tapes. As remarked earlier this form of data storage is suitable for batch type data processing which is not appropriate for production control applications. Also if the number of records to be processed during a batch run is small, the overheads involved in reading the unnecessary data can be high.

The data stored on magnetic tapes is usually designed to suit one or a limited range of applications. Engineering production data are used in a large number of company departments, for a wide range of applications. It is desirable that the data should be stored on discs in the form of a number of linked files. These linkages make it possible to represent the true data relationship between relevant data items accurately. In such a system, by following the links between related data items, it is possible to retrieve the required data without the need to process irrelevant records. Also the same basic data can be presented in a number of formats for use in different departments. It is no longer necessary to build files for use in individual departments, and the artificial departmental boundaries are not represented in files. The problem of communications between the users in a number of departments is also minimized.

Such a database in which items are cross referenced reduces the need for duplication of data. The same data are used in all the departments since there is only one data entry point. These linked files required for production applications would be referred to as the production database, and ideally should be organized in the form of:

1   Part master file.
2   Product structure file.
3   Routing file.
4   Work centre file.
5   Tool file.

Fig. 4.1 is a symbolic representation of a production database. The part master file is very dynamic when used for inventory control applications whereas the other four files contain relatively stable data. In addition it is usually necessary to create a number of temporary files in which continuously changing data are stored.

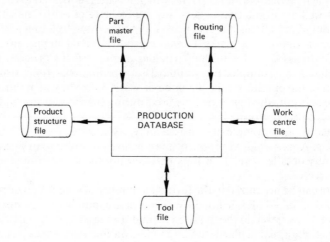

Fig. 4.1 Linked files used in a production database

Chain files used to connect records in a logical sequence, facilitate the direct retrieval of associated records. Basically the chain files consist of a series of records connected by pointers; each record contains pointers which link it to the next and previous record in the series. In the routing file, for example, pointers are used to relate all the manufacturing operations carried out on a particular part.

Part Master File

In an integrated production control system the part master file is one of the largest files, containing a vast amount of information. Each assembly, sub-assembly, component as well as raw material item is treated as a unique item. In addition it is necessary to store information which links records in the part master file to relevant records in the product structure and routing files. The data stored in the part master file originates in a number of company departments, for example engineering design, purchasing, production control, sales, production engineering, stores and operations research.

The detailed information included in a part master file would vary from one organization to another depending upon applications as well as the policy adopted by the company concerned. For example some companies might wish to use statistical forecasting techniques which requires the storage of relevant smoothing factors while other companies, such as shipbuilders, might decide that statistical forecasting is not suitable for their products. Consideration should be given to the inclusion of the following data items in the part master file:

Part Number-which uniquely identifies the item.

Part Description-which describes the part or raw material as the case may be.

Engineering Drawing Number.

Alternative Part Number-which can be used if the main part number is not in stock. This may not be applicable in all cases.

Part Type field will indicate whether the item is manufactured or purchased or both.

Part Classification according to ABC analysis.

Stores Location, i.e. the bin number in which the item is stocked.

Unit of Measure, which refers to the weight or length for raw materials and a numerical value for components.

Part Weight will show the weight of one unit.

Part Code to show whether the item is an assembly, sub-assembly, component, or raw material.

Low Level Code, i.e. the lowest level at which the item is used in the product structure file.

Unit Cost, which would represent the cost of one unit of the item whether purchased or manufactured.

Unit selling price of the item.

Total inventory on hand.

Demand for the part during current period.

Quantity received into stock during current period.

Quantity issued from stock during current period.

Quantity allocated to orders.

Backordered quantity.

Last Transaction Date, which refers to the date on which the last transaction involving this item was processed.

Usage Date 1, i.e. the date from which part should be used.

Usage Date 2, i.e. the date after which part should not be used.

Manufacturing lead time.

Purchase lead time.

Lead time code to show which of the two codes above should normally be used.

Safety lead time factor.

Supplier code number.

Minimum order quantity.

Maximum order quantity.

Order multiple factor to be used with items for which the order quantity must be rounded up so that it is a multiple of the minimum order quantity.

Minimum number of time periods for which the demand should be accumulated to create a manufacturing or purchase order.

Safety stock.

Re-order level at which a new order should be placed, which is the sum of the safety stock and the quantity used up during the lead time for the item.

Shrinkage factor used to determine the actual order quantity in order to take account of expected scrap and other losses.

Quantity on order for production items.

Quantity on order for purchased items.

Requirements for the item over future periods. The number of future periods for which requirements are accumulated and recorded will vary.

Planned orders by time period and quantity.

Released orders by time period and quantity.

Cumulative quantity used in a number of specified time periods.

The number of periods over which the cumulative quantity above was actually used up.

Date on which a physical inventory count was carried out.

Date of next physical inventory count.

For costing applications it would be necessary to store additional information. For each item assembled at a number of levels the stored data should show, for individual levels of assembly;

Standard material costs.
Actual material costs.
Standard labour costs.
Actual labour costs.
Standard overhead costs.
Actual overhead costs.
Standard total cost.
Actual total cost.

In addition fields should be allocated to the following costs up to each level of assembly.

Cumulative standard material costs
Cumulative actual material costs
Cumulative standard labour costs
Cumulative actual labour costs
Cumulative standard overhead costs
Cumulative actual overhead costs
Cumulative standard total cost
Cumulative actual total cost

Forecasting code to indicate if statistical forecasts of future demand should be made.
Model code to show the forecasting model to be used.

Alpha factor.
Beta factor.
Gamma factor. } defined in chapter 6.
Delta factor.
Mean absolute deviation.
Number of periods for which forecasts should be made.
Forecasts by time period.
Sum of squares of the errors.
Calculated value of tracking signal.
Pre-defined value of tracking signal with which the calculated value should be compared.
Number of times the pre-defined value of the tracking signal has been exceeded.

It might also be desirable to keep a record of the historical product demand data for a specified number of periods.

In addition the part master file should contain the necessary information for linking records to other files in the database, e.g. the product structure file, routing file, etc. This information is in the form of pointers. Each part number will have a pointer which indicates the higher level assembly in which the part is used and also a pointer to the address of the first component in the product structure. Another field will contain the address of the next part number with the same low level code as this part. Within the product structure file all the other relationships are created. Similarly the part master file will contain a pointer to the first manufacturing operation, carried out on the part, recorded in the routing file. Another field in the part master file is used to keep the disc address of the part record so that it can be retrieved and used by other related files. Pointers are also used to link this file to purchase order and production order files.

Product Structure File

This file defines the structure of a product and is usually linked to the part master file. In the file is contained one record for each component used in a finished item, assembly, sub-assembly, or other components and raw materials at the lowest level (i.e. components which do not have any constituent piece parts).

The low level code associated with an item indicates the lowest level at which an item is used in all the product structures. Every time a component is added or deleted it is necessary to update the low level code. The use of low level code reduces the amount of processing which must be carried out during summarized explosion and implosion calculations.

Relationships between items are established by means of pointers. One of the pointers links, for example, a sub-assembly to the assembly at the higher level; simultaneously a pointer is attached to one of the constituent components at the lower level. Raw materials are recorded only once since there are no lower level items. Since an assembly will consist of a number of components and a component may be used in more than one sub-assembly or assembly it is necessary to use a system in which a large number of product structure relationships can be established by means of the following pointers.

1   A pointer to relate the item to the part master file.
2   A pointer to relate the item to the record of the parent (higher level) item.
3   A pointer to the address of the next component in the structure of the parent item.
4   A pointer to the address of another (last) assembly in which this component is used.
5   A pointer to the address of the next assembly in which this component is used.
6   A pointer to one of the constituent parts of this item.

Once these pointers have been established it is possible to retrieve the stored information in a number of different ways. The product structure file forms the basis of Bill of Materials Processing (BOMP) programs, discussed in detail in the next chapter. The following fields should also be included in the product structure file.

1   Number of components used in one unit of the higher level assembly.
2   The date from which a particular component should be used in this product structure.
3   The date after which a particular component should not be used in this product structure relationship.
4   Operation number, in the routing file, where this component is first used.

In some cases it might be considered desirable to keep records of the engineering change numbers relating to a particular item and a shrinkage factor which shows the average percentage of components scrapped during assembly. This will have the effect of increasing the gross requirements.

Whenever pointers are included in a record it is desirable to check that the correct field is being used. The necessary cross checking can be carried out by maintaining, with each pointer, a distinct number in the parent record as well as the record being referred to. While reducing the possibility of errors it also leads to increased storage space requirements.

Routing File

The routing file describes the operations which must be performed to manufacture an item. The input to this file usually originates in the production engineering department which develops and evaluates the different manufacturing techniques. The production engineering personnel will specify the most efficient method, along with any available alternative method, for manufacturing the product and will also list the sequence in which the operations should be performed. The production engineering and work study departments determine the operation times required for production planning, costing and pay-roll applications and these data are also stored on the routing file. A typical routing file should contain information such as:

1   A pointer indicating the address of the part number on the part master file.
2   Machine set up time for performing the operation.
3   Standard labour time allowed for the operation. This standard labour time may be for one unit or a standard batch size.
4   Standard machine time allowed for the operation. As with standard labour time, the standard machine time may be for one unit or a standard batch size.
5   Standard batch size.
6   Operation sequence number.
7   Brief description of the manufacturing operation.
8   Manufacturing method sheet number which gives further details of the manufacturing operation.
9   Work/machine centre number at which the operation is to be performed.
10   Machine on which the operation should be performed.
11   Tool required to perform the operation.
12   Address of the next operation in which the same tool is used.
13   Queue time before processing.
14   Queue time after processing.
15   Minimum send ahead quantity for overlapped operations.
16   Code to indicate applicability of forced split of the operation.
17   Code to indicate applicability of economic split of the operation.
18   Split factor (defined in chapter 11).
19   Shrinkage factor for the operation.
20   Address of the alternative operation.

In addition the routing file will contain a number of pointers which link the routing file to other related files. One field will contain the address of the work centre at which the operation is performed. Similar pointers indicate the addresses of the next as well as the previous operations performed on the item. Another field contains the address of the tool in the tool file. Pointers are also maintained to show the previous and following operations which are performed in the same work centre. Similarly another pointer will contain the address of the next operation on which the same tool is used.

It is often necessary to select alternative routes for the whole manufacturing process or perform a particular operation on a different machine. For instance a numerical control machine may break down or due to inadequate capacity it might be necessary to machine some items on ordinary lathes. In such cases it is desirable to store alternative manufacturing routes/operations in the routing file. Fig. 4.2 illustrates this condition schematically.

The information retrieved from the routing file shows how the particular item is to be manufactured. A suitable computer program can be used to produce the required routing documents. In the absence of a computerized routing file it will

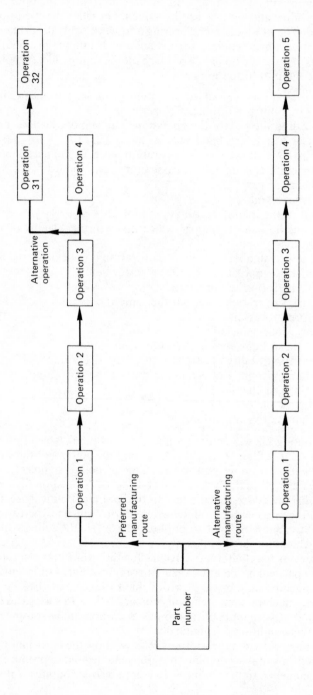

Fig. 4.2 Schematic representation of preferred and alternative manufacturing routes

be necessary to prepare manually all the required routing documents and in addition it will not be possible to carry out accurate capacity planning and shop loading calculations. Capacity planning and shop loading functions require this detailed information along with the information about the capacity of the work centres at which the operations are to be performed.

Although the routing file is relatively static, it will be necessary to maintain and update the file to reflect the changed routes.

Work Centre File

This file contains permanent information for each work centre in the company. A work centre may be an individual machine or a group of machines. It is assumed that all the machines within a work centre are interchangeable and can be used to perform a given task. Usually each work centre is identified by a unique number. This information is referred to in the routing file and used for capacity planning and shop loading calculations. For realistic capacity planning and production scheduling applications it is essential that the information stored in this relatively stable file should be accurate and up to date.

A typical file will contain the following information, for each work centre.

1   Identification number, for example a department number in which there might be several work centres.
2   Name.
3   Description.
4   Location.
5   Normal capacity.
6   Maximum capacity.
7   Unit of measure for the capacity e.g. hours.
8   Efficiency factor for the work centre.
9   Number of machines.
10   Average value of transportation time to other work centres.
11   Set up rate.
12   Machine rate.
13   Labour rate.
14   Overhead rate.
15   Alternative work centre at which similar work can be performed, albeit at a higher cost.
16   Pointer to the alternative work centre.
  For work centres containing several machines, the following additional data will be required.
17   Machine number.
18   Machine capacity.
19   Machine description.
20   Maintenance data relating to the machine, e.g. date of last maintenance, preventative maintenance time intervals, machine run time since last maintenance etc.

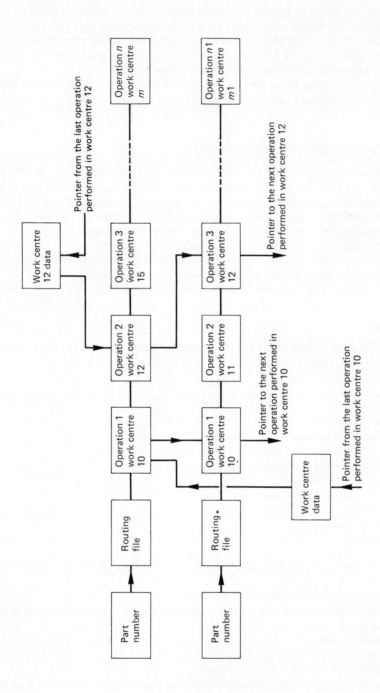

Fig. 4.3 Chaining of all the operations performed in a particular work centre

In addition the file will contain pointers to show the relative address of this particular work centre and the first operation for which this work centre is used. Thus all the operations for which a particular work centre is used can be chained together thereby making it possible to assess easily the effect of any cost changes. Fig. 4.3 illustrates the use of this technique schematically.

It would also be necessary to keep a record, by time period, of the labour and machine hours allocated to the work already planned or released to the shop-floor, along with the spare capacity which can be used for further capacity planning and shop loading calculations.

Tool File

The tool file is used in conjunction with the routing file and contains all the information necessary to identify and describe the tool uniquely. Fields represented on the tool file include:

1   Unique tool identification number.
2   Address used to access the tool number via the routing file.
3   Name of the tool.
4   Description of the tool.
5   Tool drawing number.
6   Address of the first operation, on the routing file, where this tool is used.
7   A field indicating the present status of the tool, i.e. whether it is in use, is being repaired or is available for use.
8   Work centre in which tool is being used.
9   Tool storage location.
10   Alternative tool which may be used.
11   Address of the alternative tool.
12   Estimated tool life.
13   Accumulated period for which tool has been used.
14   Unit of measure for use with fields 12 and 13.
15   Date of last tool repair operation.

The information in fields 12 to 15 can be used to inspect and repair the tool at regular intervals.

Features of a Production Database

Once created the production database is used for a wide range of applications and the decisions made are dependent on the accuracy of the data included in the database. It is essential that the input data should be carefully edited before they update the database. The data validation can take a number of forms. For example, if a new record is being added to a file then a check can be carried out to ensure that the record does not already exist on a file. Similarly in the case of changes to or deletion of records, the input validation program can check that the record already exists on file. Other input validations include checks to ensure that all the required fields are present on the transaction, numeric fields contain only numeric data and any fields used for codes contain valid values.

It is impossible to build a system in which all the possible errors can be detected; they do creep in from time to time. Therefore an audit trail of all the transactions should be kept so that the errors can be corrected at a later stage. An audit trail also helps in the re-creation of the database in the event of a system breakdown. In batch mode input transactions should be listed and invalid data flagged. This flagging might be in the form of some special characters, for example asterisks, alongside that transaction and the field containing errors. An alternative and better approach would be to list all invalid transactions in the form of an exception report so that they can be corrected quickly.

If one data entry program is used in the batch mode to input all the transactions data, followed by the execution of a program which can split this data, according to specified codes, then the necessity for manually separating the input cards for different types of transactions is eliminated. Subsequently a sort routine can be used to sequence the separated transaction files in the required order.

For a system which is used mainly in batch mode, information about the total number of transactions input to a file along with the numbers of valid and invalid transactions should be kept, so that any necessary manual data reconciliation can be carried out. If a single program is used to input all the data, which is then split for use with individual files, then a similar record of the number of transactions input to these separate files should be kept.

Data relating to all the invalid transactions should be stored on a separate file. As correct transactions are entered the corresponding wrong data stored on the invalid transactions file are deleted. If no corrections are entered within a specified period, for example before the next batch run, then an exception report can be generated for the attention of the users and management.

Programs should be available for the reorganization of the whole of the database or individual files. Periodic reorganization is necessary because any additional records input to the datafiles are stored in different areas of the disc. Similarly records are often deleted from the files and for efficient utilization of the disc space it is necessary to reorganize the database. Since there are links between different files it is necessary to ensure that the file reorganization programs take account of these links. For example the part master file is linked to the routing file and the product structure file. Consequently these chains also have to be updated.

Whenever any changes are made to the product structure file a list of the structure of all the products affected by that change should be produced. Product structure changes usually initiated by the engineering design department are in the form of addition or deletion of components, or replacement of one component by another. Such a listing can be of considerable value in ensuring that the design department has not overlooked the use of the component in other end items. Similarly when the routing file is altered by the addition, deletion or replacement of a particular manufacturing operation, a list of the effect of the changes should be produced to ensure that the production engineering/design departments are fully aware of them. This is particularly necessary in the batch mode of computer operations.

Programs should be available for retrieving information stored on the product structure file in different formats such as single level explosion, indented explosion, summarized explosion, single level implosion. These are further discussed in the next chapter. Similar programs should be available to retrieve details of the operations required to manufacture parts, sub-assemblies or end items from the routing file. Details of the various operations performed in a work centre can be retrieved from the work centre file by means of a suitable application program. Facilities should also be available to list complete or selected sections of files.

When a production database is being created the requirements of functional areas other than the production and engineering departments should be taken into account. It might be desirable to build an overall company database in which financial, commercial and personnel data are integrated with the production data.

The task of creating, organizing and reorganizing the datebase and retrieving the data in different formats can be eased by using proprietary database management system software.

As remarked earlier it is best to create and use the production database in the on-line mode. Changes can be directly input to the system and the effect of engineering modifications simulated. In the case of any amendments to the routing file, the existing route can be examined and information relating to the amended route and the date on which it becomes effective can be input to the system. On the effective date the previous route can be automatically deleted from the file. Similarly in the case of engineering design work the existence of similar parts can be immediately identified thereby eliminating wasted efforts in the design and production departments. Bill of materials data for new products can be directly input to the system and new product structure relationships created. The effect of any alterations to the existing product structure and routing records can be examined immediately and any errors avoided.

On-line data validation procedures can be used to check that the data entering the system are accurate. In the event of any errors being encountered the terminal operator can be informed immediately and requested to enter correct data. This requirement to ensure the accuracy and integrity of the datebase cannot be overemphasized.

D

5
Bill of Materials Processing

Introduction

It is the usual practice, in most engineering companies, to show the bill of materials on the top assembly drawing. While this document shows the components and their respective quantities used in the assembly, the engineering drawing does not indicate the hierarchical relationships between raw materials or components at the lowest level and the sub-assemblies or assemblies and the way in which these components should be assembled to form the final product. Also the top engineering drawing only indicates the use of components in a particular assembly. Frequently a component can be used in a number of sub-assemblies or assemblies. With a manual system it is an ardous task to determine the finished products in which a particular component or sub-assembly is used unless a separate record of the usage of components is maintained.

Foundations of Bill of Materials Processing

Manufacturing organizations can gain considerable benefits by using 'Bill of Materials Processing (BOMP)' programs. The prime function of a BOMP system is to create and maintain the product structure records and establish relationships between items at various hierarchical levels of a product. The relationships between individual items, with unique identification numbers, are established in the product structure file.

The product structure also defines the way in which individual components are assembled in a number of stages. The quantities of individual components required to produce one unit of a given product are also specified in this file. Such a system can be used to identify the materials, component parts and sub-assemblies which are used in an assembly, finished products or spare parts. The BOMP system can also be used to identify the assemblies or products in which a given raw material, component or sub-assembly is used.

The product structure data usually originates in the design department in the form of design drawings and the list of components and the quantities associated with a particular design. A block diagram is used to represent the product structure data illustrating the linkages between materials, components and assemblies at different levels within the product. Fig. 5.1 shows a typical product structure.

This product structure is often referred to as a family tree structure. This representation is easy to understand and the various stages of the assembly process can be easily identified and comprehended. In this case it can be seen that Assembly A is made up of 2 units of purchased part B, 1 unit of purchased part D, along with 2 units of sub-assembly C. In turn 1 unit of sub-assembly C is

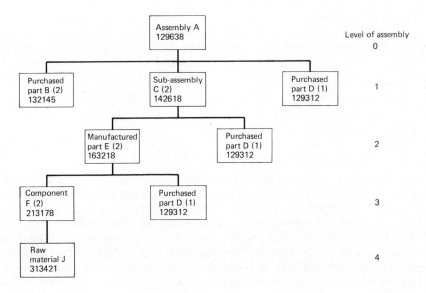

Fig. 5.1 A typical product structure tree

assembled from 2 units of manufactured part E and 1 unit of purchased part D. The product structure also shows that manufactured part E comprises 1 unit of purchased part D and 2 units of component F and that material J is used for its manufacture. Parts explosion can be simplified by allocating a code to each assembly level; the level of assembly is also shown in Fig. 5.1. In this particular type of system, level '0' is allocated to the end products. Sub-assemblies and purchased parts directly used in the assembly are allocated level 1. This process is continued until raw materials at the lowest level of the product structure are reached. In practice a component may be used at a number of different levels in product structure relationships. To avoid the difficulties which would be encountered during the bill of materials processing, the low level code associated with an item is the lowest level at which it is used anywhere in the product structure file. For example, a component may be used directly (level 1) as well as at a lower level (say level 4) in one assembly, and in another assembly it might be used at the 6th level. In the product structure file the low level code allocated to the particular component would have a value of 6 and it would be processed only after the 6th level has been reached so that the earliest requirements for that component can be catered for. The allocation of level codes makes it possible to compute the total requirements for individual components easily when a number of end items are exploded. This type of 'Top-Justified' system in which finished products are allocated level '0' is very commonly used. In other systems lowest level code is allocated to items without any constituent piece parts. Thus raw materials and purchased parts are at the lowest level which might have a value of zero or one. The level of any particular item is calculated by adding one to the constituent component with the highest level number. Therefore, by definition, in such a 'Bottom-Justified' system the end items are identified as those with highest level assembly code in the product structure.

Once the relationships between items at various levels of a product structure have been established the required data can be retrieved, by means of suitable retrieval programs, in the formats required by the user. Frequently used formats are:

1   Single level explosion
2   Indented explosion
3   Summarized explosion
4   Single level implosion (where used lists)
5   Indented implosion
6   Summarized implosion

It is necessary to have programs for retrieving the product structure data in a number of formats because a single format is not suitable for all the applications in a number of different departments. For example, the single level explosion lists the components and sub-assemblies directly used in a particular assembly.

The single level bill of material is useful for performing level by level net requirements planning, showing how an item is directly manufactured, and preparing material pick-up lists. If one of the constituent items of an assembly is a sub-assembly with constituent components, then a separate single level bill of material for the sub-assembly can be retrieved. In the case of an item with a multi-level structure only a small part of the total structure is shown on this bill. The single level bill of material for the assembly A in the simple product structure illustrated in Fig. 5.1, will be as shown in Fig. 5.2.

SINGLE LEVEL EXPLOSION

ASSEMBLY A
PART NO. 129638

PART NO.	PART DESCRIPTION	QUANTITY/UNIT
132145	Purchased part B	2
142618	Sub-assembly C	2
129312	Purchased part D	1

Fig. 5.2 Single level explosion for Assembly A (Fig. 5.1)

The single level bill of materials data, shown above, while easy to maintain is not useful for showing the sequence in which end products are assembled. Indented bill of materials are used to show the detailed structure of the end products or major assemblies and provide design departments with up to date documentation relating to the product structure. Indented parts lists are useful for developing product lead times and preparing spare parts catalogues. All parts directly and indirectly used in the higher level item are listed on the indented bill of materials which also indicates how the various components are used in the higher level item. A component used at different levels within a product structure is repeated at each level of its use. As a result the indented bill of materials often becomes a very long list. With reference to the product structure illustrated in Fig. 5.1 the indented bill of materials is as shown in Fig. 5.3.

INDENTED BILL OF MATERIALS

ASSEMBLY A
PART NO. 129638

LEVEL CODE	PART NO.	PART DESCRIPTION	QUANTITY/UNIT
*1	132415	Purchased part B	2
*1	142618	Sub-assembly C	2
**2	163218	Manufactured part E	2
***3	213178	Component F	2
****4	313421	Raw material J (Bar)	12 cm
***3	129312	Purchased part D	1
**2	129312	Purchased part D	1
*1	129312	Purchased part D	1

Fig. 5.3 Indented bill of materials for Assembly A (Fig. 5.1)

The summarized bill of materials shows the components, along with their quantities, required for assembling a given product. Each component is listed only once, at the lowest level at which it is used and the total quantity of the component used in that assembly is shown. The summarized parts list is useful for gross requirements planning and product costing applications. The summarized bill of materials for the assembly A (Fig. 5.1) is shown in Fig. 5.4.

SUMMARIZED BILL OF MATERIALS

ASSEMBLY A
PART NO. 129638

LEVEL CODE	PART NO.	PART DESCRIPTION	QUANTITY/UNIT
*1	132145	Purchased part B	2
*1	142618	Sub-assembly C	2
**2	163218	Manufactured part E	4
***3	213178	Component F	8
***3	129312	Purchased part D	7
****4	313412	Raw material J (Bar)	12 cm

Fig. 5.4 Summarized bill of materials for Assembly A (Fig. 5.1)

It is often necessary to retrieve information relating to the assemblies or finished products in which a particular item is used. Alternatively it may be desirable to determine the level or levels at which a particular item is used in a product structure. These 'where used' reports can be produced in formats similar to those for bill of materials data, i.e. single level 'where used' list, indented 'where used' list, and summarized 'where used' list.

A single level 'where used' list shows the use of an item, along with the relevant quantity, on all the sub-assemblies or assemblies at the next higher level. This information can be used for the allocation of components and for analysing the effect of any proposed engineering changes. In addition part interchangeability may also be examined. A single level 'where used' list is also useful when some fault in a component is discovered and the component must be replaced. Fig. 5.5 shows a single level parts implosion report relating to the use of purchased part D in Assembly A.

SINGLE LEVEL IMPLOSION REPORT

PURCHASED PART D
PART NO. 129312

PART NO.	PART DESCRIPTION	QUANTITY
129638	Assembly A	1
142618	Sub-assembly C	1
163218	Manufactured part E	1

Fig. 5.5 Single level 'where used' parts list for purchased part D

If the purchased part D is used in other assemblies then the direct use will be listed on the single level implosion report.

The indented 'where used' list shows not only the assembly or sub-assembly at the higher level in which a component is directly used, but also the next higher level assembly or finished item in which the sub-assembly is directly used. This process continues through all assembly levels until the end item is reached. Thus in the case of parts used at a number of levels, the indented 'where used' list repeats the end item a number of times within the particular product structure. This report identifies the items which are directly or indirectly affected by any engineering changes and helps the design department in making decisions about whether or not to go ahead with the change.

The summarized 'where used' list shows the higher level assemblies in which a part is used. However in this report each of the end items is listed only once along with the total quantity of that part used in one unit of the parent assembly. This report can be used in production planning and control work and for allocating components. The summarized implosion report helps in analysing the impact of the shortage of components on production schedules. The effect of any increase in the cost of particular items can be quickly assessed. These 'where used' lists can also be used for value analysis and cost reduction exercises.

Depending upon the contents of the part master and product structures files the bill of materials and 'where used' lists might include additional information such as whether the item is manufactured or purchased, the date after which a particular component should be used, engineering drawing number, and lead time.

BOMP Programs

Programs used for retrieving the bill of materials and product structure data employ the part master file records for obtaining access to data relating to that part, and the product structure file for the retrieval of information relating to the relationships between various items. Thus in the bill of materials type processing the input data indicates the part number of the assembly. The part master record for the assembly contains the disc address at which the product structure relationships for the assembly are stored. By following the chain of all the components used in the assembly, the bill of materials list can be prepared. Additional information pertaining to individual components can be retrieved from the part master file. Each component listed in the product structure file contains a field in which the disc address of the component in the part master file is recorded. Let us look more closely at the routines used for single level, indented, and summarized explosions.

Single Level Explosion

In parts explosion routines the assembly–component chain is followed. The disc address of the first component used in an assembly is retrieved from the part master file, where the detailed data relating to this component is stored. One of the fields in the product structure file contains the disc address of the component record on the part master file, and the data stored in this field can be used to gain access to the required detailed information. Once this information has been retrieved and stored on a temporary working file, the disc address of the next component used in the assembly is obtained. This process is repeated until information relating to all the components used in the assembly has been retrieved. The information stored on the temporary working file can then be displayed on a visual display unit or printed out as required.

Indented Explosion

The indented bill of materials list is prepared by following the same basic procedure as used for single level explosion. The only difference is that if a constituent item in a product structure is an assembly in its own right, then it would be necessary to branch out and explode the lower level assembly before continuing with the explosion of the parent assembly. This branching out procedure minimizes the amount of data processing which has to be carried out. Explosion from one level to the next has priority over the explosion within the same level. Thus for preparing indented bill of material lists the explosion process is repeated at a number of levels within a given product structure.

Summarized Explosion

The basic procedure used for preparing summarized bill of materials lists is similar to the one used for single level bill of materials. For summarized explosion it is also necessary to keep a record of the part number, its low level code, and the required quantity every time a new item is encountered. If the item has previously been recorded then the quantity required at a particular level is added to the previous quantity. Once the parent assembly has been exploded, the part whose low level code is immediately greater than the low level code of the parent assembly is further exploded. By repeating this process until no further explosions are possible the summarized bill of materials list can be prepared.

Implosion programs

Part master and product structure files are also used in implosion 'where used' routines and the part 'where used' pointers are followed. The disc address of one of the higher level assemblies, in which a particular component is used, is retrieved from the part master file and used to initiate the retrieval of data from the product structure and part master files. Within the product structure file a chain links the records of all the assemblies in which a particular part is directly used. By processing this data in sequence all the information relating to the direct and indirect usage, at various assembly levels, of a given part can be quickly and easily retrieved. The routines used for single level, indented, and summarized implosions are as follows.

Single Level Implosion

By reading the part master record of the part to be imploded the disc address, on the product structure file, of the first assembly in which the part is used can be retrieved. One of the fields in the product structure file contains the disc address of the assembly record on the part master file, and the value stored in this field is used to access the detailed information about that assembly. Once this information has been retrieved and stored on a temporary working file the disc address of the next assembly, in which this component is used, can be obtained. By repeating this procedure for all the assemblies in which the component is directly used the single level 'where used' list is prepared.

Indented Implosion

Indented implosion procedure is similar to that used for single level implosion. The data processing requirements can be minimized by giving priority to the vertical implosion. Thus if the assembly, in which a component is directly used, is a constituent of a higher level assembly then the implosion of the higher level assembly should be carried out until the end item is reached when the implosion of the part in directly used assemblies can be continued. The complete indented 'where used' list is prepared by repeating this procedure for all implosions.

Summarized Implosion

Summarized implosion consists of a number of single level implosions. As with summarized explosions a record of the part number, its common low level code, and quantity required, is kept to ensure that the part is not repeated. If the item has previously been recorded then the quantity required at a particular level is added to the previous quantity. When the direct usages of the item have been determined the implosion process is continued for the assembly whose low level code is immediately less than the low level code of the previous assembly or component. By repeating this process until all the end items have been reached the summarized 'where used' list can be prepared.

Considerations in the Development of BOMP Systems

It is desirable that the product structure data should be loaded in the form of single level assembly relationships. By following the pointers between different components and assemblies, the use of a component or an assembly in any number of higher level assemblies can be easily determined and listed, using indented or summarized implosion programs. Also the need to repeat an item every time it is used in a higher level assembly is eliminated thereby making a better utilization of the available mass storage space.

In some engineering organizations, for example aerospace companies, the product structure data are very volatile. In other companies, also, it is frequently necessary to alter the product structure by adding, deleting, or changing some components. Obsolete items have to be removed from the files which would otherwise grow to unmanageable proportions with a consequent waste of storage space. The dumping of files from one storage device to another will take a long time. It is, therefore, desirable that the product structure and part master files should be continuously updated, and reorganized at regular intervals.

An on-line real-time terminal based system can offer considerable assistance in the creation and maintenance of bill of materials data. This is particularly true when it is necessary to create a new bill of materials from an existing one. The existing bill of materials can be displayed on the user terminal, and the user can delete some of the component-assembly relationships, and add new ones. The user can also examine the effect of these amendments on other components or assemblies. This procedure reduces the time required to create a new bill of materials.The possibility of errors of oversight is also minimized since the engineer can quickly assess the impact of proposed changes. In some cases the date on which an engineering change becomes effective may not be the same as the date on which a new product structure relationship is input to the system. For example, an engineering change may not become effective until the quantity in stock of the 'old' item has been used up. In such cases it is necessary to store the 'old' and 'new' product structure relationships until the change becomes effective. This can be achieved by associating a date with the old and new product relationships; on the specified date the new product structure relationships will become effective and the old relationship may be deleted. This should be taken into account in the development of application programs. If these changes are not reflected in the appropriate files then the old part will continue to be made, and the final product may not meet revised specifications. The addition of the new product structure relationship before the effective date ensures that old parts or raw materials required for the manufacture of old parts are not purchased.

6
Forecasting Systems

Introduction

In order to maintain a smooth flow of work in the production shops and to ensure that the product demand is met, the management of most manufacturing companies have to make estimates of the future product demand. These estimates might be based on an educated guess, close liaison with the customers, or the use of mathematical models. Forecasts are necessary for the preparation of a master production schedule which shows the expected requirements for finished products as well as spare parts. They are also required to determine the stock levels of particular products and raw materials which should be maintained. Forecasts of other factors such as the machine breakdown period, labour efficiency, requirements for tools, are also necessary.

If enough items are not available when required, then sales might be lost along with customer goodwill. This makes it necessary to base some of the manufacturing plans on the estimated demand for the product. In rare cases when the manufacturer has a long order book, sales forecasts might not be required, but at some stage or other the use of forecasting techniques is inevitable, e.g. with regard to the stock levels and scrap percentages. In the present context forecasting refers to the projection over future time periods of historical time-series of numerical values collected over a period of time. The historical data relating to the process under study is usually collected at fixed time intervals, e.g. weekly, monthly, quarterly, etc.

In most organizations the customer demand fluctuates and it is extremely difficult, if not impossible, to predict correctly the future level of demand. It is, perhaps, no exaggeration to say that if on a number of consecutive intervals the forecasts exactly match the demand for the product then this match is purely coincidental. However, the calculation of the forecasting error in itself helps to maintain a correct level of safety stock, over a period, and therefore cushion the effect of variations in the demand level. There has been considerable progress in the development of efficient forecasting techniques, but none of the currently available techniques can produce absolutely accurate forecasts of future events.

Most manufacturing companies do not prepare their forecasts in a formalized manner. They are frequently based on guesswork or intuition, without an adequate analysis of the past data. With such informal systems there is a general tendency to over-react and increase the safety stock so that the effect of subjective forecasting errors is minimized. Even moderate increases in safety stock levels, when spread over thousands of items, result in a large increase in the level of investment in the inventory.

Forecasting Methods

A large number of forecasting methods is available, but no single one is suitable in all circumstances. However, some methods are more efficient than others. Intrinsic forecasting techniques are based on the use of historic demand data relating to the product itself. Traditionally management has used moving average forecasting techniques to predict the level of future demand. This particular technique is suitable for data which are either stable or change very slowly. Basically the method gives equal weight to all the past data and the future demand is calculated by averaging the past demand. The method is not suitable for data which either exhibit trends or are fast moving. The exponential smoothing technique, which attaches more weight to recent data, is more reliable than the moving average method. The weight attached to the latest and past data points can be altered by modifying the smoothing factor.

Extrinsic forecasting models make use of data relating to some external factor, usually referred to as the leading indicator, in addition to the product demand data. Typical leading indicators are gross national income per capita, sale of basic materials such as steel, the number of new housing starts, etc. The cost of developing extrinsic models which show causal relationship between the product demand data and leading indicator data, is often high. Such forecasting techniques are not suitable for individual minor items.

Forecasts can also be based on the judgment of company personnel who have specialist knowledge of the market for the product. For example, the opinion of senior managers and sales and marketing personnel can be solicited and weighed. The major problem lies in consolidating these opinions. Often more weight is attached to the opinions of senior managers, because of their rank or position, even though they may not have as much knowledge of the market as the sales

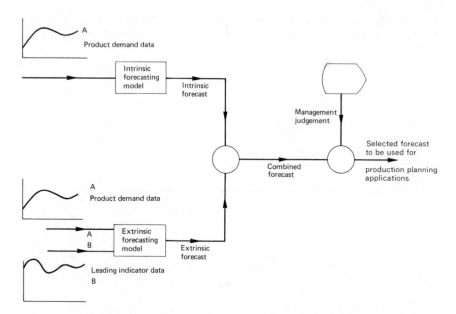

Fig. 6.1 Development of forecasts required for planning applications

personnel. Nevertheless it might be necessary to use this particular forecasting method if enough historical data does not exist, as for example when a new product is introduced.

The best forecasts can be prepared by combining intrinsic and extrinsic forecasts, and modifying the combined forecast in the light of human judgment about factors not included in the intrinsic and extrinsic models. Fig. 6.1 shows a symbolic representation of such a forecasting method.

The majority of intrinsic forecasting techniques are relatively simple and can be used by anyone, following some study of the concepts of statistical forecasting. It is important that people responsible for implementing the results of statistical forecasting techniques should have an understanding of the assumptions and details of the techniques used. Some of the frequently used techniques are now discussed in more detail.

Intrinsic Forecasting Models

A forecasting model is a mathematical equation which describes the historical data. Forecasts are produced from it by weighting the various numerical values, in the time-series, in a specified manner. In some models equal weight is given to all the past data while in other models more weight is attached to the recent data. Moving average and exponential smoothing models are examples of these two types of models. The different intrinsic models, commonly used in practice, are as follows.

(1) Constant model

The simplest forecasting model is one in which the sales fluctuate around a constant level. Fig. 6.2 shows the graphical representation of such a model.

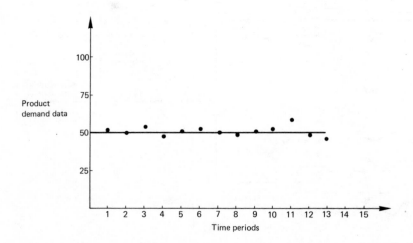

Fig. 6.2 Product demand data which can be represented by a constant model

Mathematically the model can be expressed as

$$Y = A + E \tag{1}$$

where Y = Product demand
A = Constant average
E = Forecasting error

From Fig. 6.2 it can be seen that the product demand varies around a level of 50 units. Therefore the forecast demand for future periods is the average value of the product demand for the previous periods. The forecasting error is usually measured in terms of the mean absolute deviation which can be calculated by averaging, over a specified time period, the absolute deviation of the forecast from the actual sales level.

(2) Trend models

In practice a constant model often fails to reflect the historical sales level or stock usage. The demand level might be increasing as for example when a new product is introduced. Alternatively as a product comes to the end of its life its demand level will fall. A constant model does not correct the forecasts to take account of these increasing or decreasing trends. A trend is present if the time-series data increases or decreases steadily over a number of consecutive data collection time intervals. Fig. 6.3 shows a product with increasing demand pattern.

From Fig. 6.3 it is seen that although the product demand can be represented by a straight line, this straight line has a positive slope. In this case the product demand Y can be described by the equation:

Product demand = Constant level + Slope of the line × Time period + Forecasting error

$$Y = A + BX + E \tag{2}$$

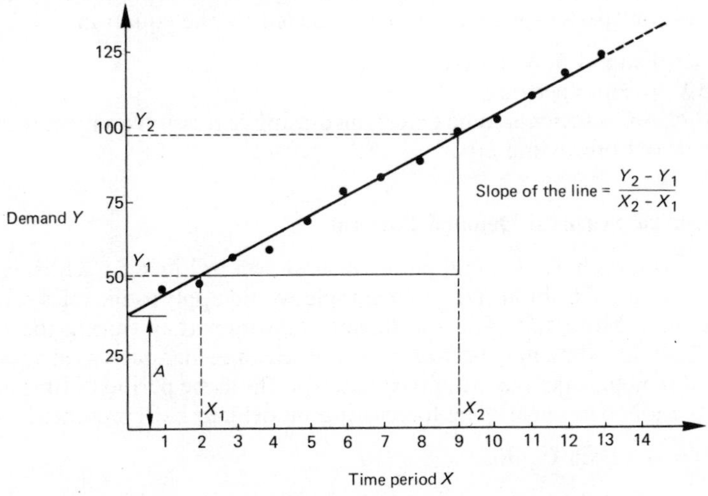

Fig. 6.3 Product demand data with an increasing trend

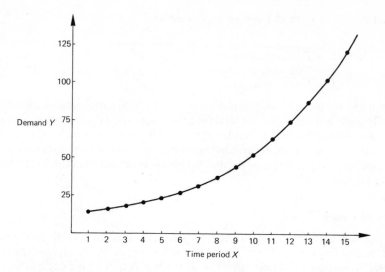

Fig. 6.4 Demand data represented by a quadratic equation

The equation can be fitted using regression analysis techniques.

The model above represents a situation in which the product demand is increasing at a constant rate. Therefore the model results in a forecast which has been corrected for the linear trend by increasing (or decreasing in the case of a product with a linearly decreasing demand pattern) the sales forecast by a constant quantity for each time period. In a computer-based forecasting system, which uses such a model, the system should be capable of calculating the base level, i.e. the sales level for the initialized time period.

The model in equation 2 does not take account of the circumstances in which the product demand is increasing or decreasing at a variable rate. Fig. 6.4 shows such a demand pattern which may be modelled by the equation:

$$\text{Product demand } Y = A + BX + CX^2 + E \qquad (3)$$
where X = Time period
A, B, C = Coefficients to be determined using regression analysis techniques
E = Forecasting error

(3) Cyclical or Seasonal Demand Patterns

Many products have demand patterns, as shown in Fig. 6.5a, which vary in a cyclical or seasonal fashion. This, for example, would apply to the sales of made to measure suits. More suits are sold during the winter than during the summer period. A demand data may be said to exhibit seasonal fluctuations in a year if the peaks and troughs take place approximately at the same period of time as in the past or the following year. The forecasting model can be represented by:

$$Y = A + B \sin X + E \qquad (4)$$

A variant of this type of model is one in which the demand varies in a seasonal fashion along with an increasing or decreasing trend. These conditions are catered for by using a model which can correct the sales forecast by taking account of the trends and seasonal factors. Fig. 6.5b shows a demand pattern with seasonal and trend variations which may be modelled by the equation:

$$Y = A + BX + C \sin X + E \tag{5}$$

Almost all patterns of demand broadly fall into one of these categories. The suitability of a particular model for representing a given time-series is usually determined by using regression analysis techniques which essentially consist of

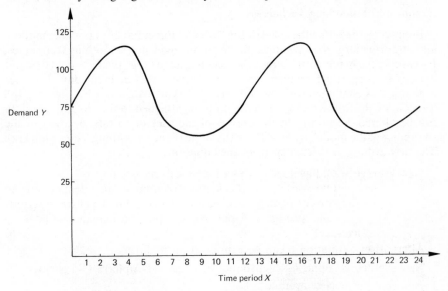

Fig. 6.5a Demand data with seasonal fluctuations

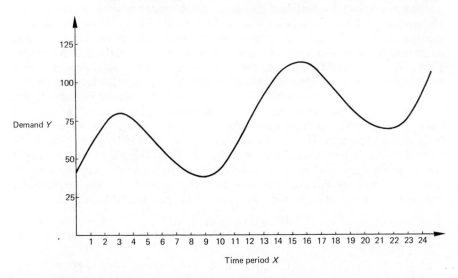

Fig. 6.5b Demand data with seasonal fluctuations and an increasing trend

fitting a curve to the given data. The choice amongst alternative models is made by calculating the sum of squares of the errors between the actual demand and the data fitted by the selected equation. In practice the choice can be difficult as the number of alternative models used to represent a time-series may be large.

The application of these techniques also requires the storage of demand data for many periods. If the number of items involved is large this might require excessive storage space.

Exponential Smoothing Techniques

In practice the difficulties of selecting from a large number of possible models, and of storing large data volumes can be minimized using smoothing techniques. Essentially the exponential smoothing techniques average the historical demand pattern in a specified manner and as the new actual demand data becomes available it is combined with the old average to calculate a new average and the forecast for the future period. The weight to be attached to the old and new data can be altered, according to the desires of the user or to take account of the suitability of the model, by modifying the value of the smoothing factor alpha (α). The new average is calculated from the equation:

New average = Old average + Alpha factor × Forecasting error
= Old average + Alpha factor × (Actual product demand during the current period − Forecast of demand for the current period)
= Old average + Alpha × (New product demand − Old average)   (6)

Alternatively, equation 6 can be expressed in the form:

New average = Alpha × New product demand + (1-Alpha) × Old average   (7)
New forecast = New average

Thus it is only necessary to store the old average along with the Alpha factor. As the new product demand data becomes available it can be used to calculate the new average and hence the new forecast. At the end of each period the last new average becomes the old average and the previous value of the old average is discarded. It can be seen from equation 7 that as the value of the Alpha factor is increased, more weight is given to the demand data for the latest period, with a proportional decrease in the weight attached to the old data. A large value of α could result in a situation in which the sales forecasts, for future time periods, are large due to a random increase in the demand during the last period. Similarly too low a value of Alpha will mean that it takes a long time to react to genuine changes in the demand pattern. There is no optimal value which can be used; it has to be adjusted in the light of available data. Values of Alpha lying between 0.1 and 0.2 are normally used. The weights to be attached to historical data vary in an exponential manner. For example, using an Alpha factor of 0.15, the weights to be attached, as current data in the 13th period becomes available, will be

Time period	Weight to be attached	Actual value
13th (current)	α	0.1500
12th (last)	$(1-\alpha)\alpha$	0.1275
11th	$(1-\alpha)(1-\alpha)\alpha$	0.1084
10th	$(1-\alpha)(1-\alpha)(1-\alpha)\alpha$	0.0921

The sum of all the weights to be attached to historical data comes to 1.000. This ensures that the new average is not biased. A graphical plot of the weights attached results in an exponential curve, hence the use of the term exponential smoothing technique. The use of this technique can cause problems when the old average value is not available and adequate historical data for calculating it does not exist. This difficulty can be overcome by using a moving average, of the actual sales, in which equal weight is attached to the data for all the periods. This moving average starts off the forecasting process. A good computer-based forecasting system should provide this option whereby moving average models are used to forecast the future sales levels, during the periods when not enough data are available. This would apply until data for the last 1/Alpha periods becomes available. Thus if the α factor had a value of 0.1, then the forecasts would be made using moving average techniques for the first 10 periods at which stage the current moving average is used for initiating the exponential smoothing forecasting technique.

Single exponential smoothing models can be used to represent a 'Constant' product demand pattern. In symbolic terms the model can be written as:

$V_T = Y_T + (1 - \alpha) V_{T-1}$
where V_T = Average value for period T
Y_T = Product demand during period T

The above exponential smoothing technique cannot be used to forecast the demand for products with linear or cyclical demand patterns. 'Double exponential smoothing' which takes into account the value by which the product sales will increase or decrease during the data sampling interval are used for trend forecasting applications. Usually it is best to calculate a smoothed value of the trend at each new time period, using the relationship:

New trend = Beta factor (new single smoothed average
    − old single smoothed average)
    + (1 − Beta factor) × Old trend

Denoting trend value at time T by b_T, we have

$$b_T = \beta(V_T - V_{T-1}) + (1 - \beta) \times b_{T-1} \qquad (8)$$

Beta (β) is the constant used to smooth out the trend values. In practice the value of Beta is half the value of the Alpha factor.

A 'one period ahead' forecast $_T\bar{Y}_{T+1}$ is given by

$$_T\bar{Y}_{T+1} = \text{New single smoothed average} + \left(\frac{\text{New trend}}{\text{Alpha factor}}\right)$$

$$= V_T + \frac{b_T}{\alpha}$$

For making forecasts more than one period ahead the trend constant b_T is added to the 'one period ahead' forecast for each further time period over which forecasts are required. Thus, for ease of computer use, the forecasting algorithm for making forecasts j periods ahead, can be written as:

$$_T\bar{Y}_{T+j} = V_T + \frac{b_T}{\alpha} + b_T(j-1)$$

or $\quad _T\bar{Y}_{T+j} = V_T + b_T\left(j - 1 + \dfrac{1}{\alpha}\right)$ $\qquad\qquad$ (9)

The sales forecast for the current period ($j = 0$) is:

$$_T\bar{Y}_T = V_T + b_T\left(-1 + \dfrac{1}{\alpha}\right)$$

$$= V_T + b_T\left(\dfrac{1-\alpha}{\alpha}\right) \qquad\qquad (10)$$

Seasonal or Cyclical Models

In the case of products with seasonal demand patterns, it is necessary to use an exponential smoothing model which takes account of the seasonal variations for the particular period. For example, if the total demand for a given product is 650 items per year, the average demand for each 4 week period is approximately 50 units. Then if the seasonal factor for a given period is 1.15 the forecasting model has to be such that the sales forecast, produced by the system, is $50 \times 1.15 = 57.5$ $= 58$ units, thereby showing a 15% increase in sales to take account of the seasonal variation. There are as many seasonal factors as the number of periods during the year and the average value of seasonal factors is 1.

A computer system used to implement such a seasonal forecasting system should be such that in addition to smoothing the sales data it also smooths and adjusts the seasonal factors. Before such a seasonal forecasting system can be implemented, it is necessary to collect about two years historical data to compute the seasonal factors accurately. The actual forecast is computed in the following stages.

$$\begin{pmatrix}\text{New single}\\ \text{smoothed}\\ \text{average}\end{pmatrix} = \begin{pmatrix}\text{Old single}\\ \text{smoothed}\\ \text{average}\end{pmatrix} + \alpha\left(\dfrac{\text{New product demand}}{\text{Old seasonal factor}}\text{ for the current period} - \begin{matrix}\text{Old single}\\ \text{smoothed}\\ \text{average}\end{matrix}\right)$$

Thus in period T of a given year:

$$V_T = V_{T-1} + \alpha\left(\dfrac{Y_T}{SF_T} - V_{T-1}\right) \qquad\qquad (11)$$

The new seasonal factor for the current period is determined using the equation:

$$SF_T = SF_{T-t_1} + \gamma\left(\dfrac{Y_T}{V_T} - SF_{T-t_1}\right) \qquad\qquad (12)$$

Where t_1 is the number of periods in the year, and Gamma (γ) is the smoothing constant for the seasonal factors. As stated earlier the average value of the seasonal factors is 1 and any increase or decrease in a single seasonal factor alters this average value, so that it is necessary to adjust all the seasonal factors. This can be accomplished by using equation 13.

Adjusted value of
seasonal factor for $=$
period

$$\frac{\text{Unadjusted value of the seasonal factor for the same period}}{\text{New Average value of all the seasonal factors arrived at after computing the new seasonal factor for period } T.}$$

(13)

Once all the new adjusted seasonal factors have been computed, the sales forecasts for future time periods may be calculated using the equation:

$$_T\bar{Y}_{T+j} = V_T \times \text{Seasonal factor for period } T+j$$

$$= V_T \times SF_j$$

(14)

In the case of demand patterns which involve trend as well as seasonal variations it is necessary to combine the procedures outlined above. Thus the new single smoothed average is first determined using equation 11. New values of the trend and seasonal factors are determined using equations 8 and 12 respectively. Equation 13 is then used to adjust the value of seasonal factors. By means of equation 10 the sales forecast for the current period is calculated. This sales forecast represents the base for calculating sales forecasts for future periods from the equation:

$$_T\bar{Y}_{T+j} = {}_T\bar{Y}_T + j \times (\text{Trend value})$$

$$\times \text{Adjusted seasonal factor for period } (T+j)$$

(15)

Forecasting Errors

Almost all forecasts contain errors. As previously stated it is only coincidental if the forecasts and actual demand levels are in agreement. The cause of many manufacturing problems is the existence of forecasting errors. Too high forecasts cause as many problems as forecasts which are too low. Too low forecasts, in relation to actual demand, result in the demand not being fully satisfied and customers changing to other suppliers. Too high forecasts lead to a build up of inventories, and the possibility that the company will acquire additional manufacturing capacity which might not really be required. The calculation of forecasting errors makes it possible to establish accurate safety stock levels used to cushion the impact of demand fluctuations. The forecast errors are usually measured in terms of the 'Mean Absolute Deviation' (MAD). The term absolute deviation refers to the difference between the forecast and the actual sales levels, and all deviations, whether positive or negative, are treated as positive for the purpose of calculating mean absolute deviations. Therefore, the mean absolute deviation is the average of all the forecasting deviations. Calculated values of MAD are used to determine the level of probability of the demand for a product lying between specified values. Assuming that at a given time period the values of the forecast and the Mean Absolute Deviation are 50 and 4 respectively, then the probability that the demand for the product in the next time period would lie between

46 and 54 is 78.8%	(Forecast $\pm$ 1 MAD)
42 and 58 is 94.5%	(Forecast $\pm$ 2 MAD)
38 and 62 is 99.2%	(Forecast $\pm$ 3 MAD)

The values of the mean absolute deviations are usually updated as new values of the demand levels are received. The new MAD is calculated using the equation:

$$\text{MAD}_T = \text{MAD}_{T-1} + \delta \, (\text{Absolute value of} \, (Y_T - {}_T\bar{Y}_T) - \text{MAD}_{T-1}) \qquad (16)$$

Delta (δ) in equation 16 is a smoothing constant. Usually it has the same value as the constant Alpha used to update the single smoothed averages. The smoothed value of MAD can then be used to calculate the confidence intervals for the forecasts. However, these confidence intervals apply only if the forecasting error is truly random and the forecasting model used is the correct one. For example, if a linear trend model is used to calculate forecasts for data with seasonal variations, then the forecasting errors will be biased and the confidence intervals will no longer apply. A tracking signal used to measure this possible bias may be calculated from the equation:

$$\text{Tracking signal} = \frac{\text{Sum of forecasting errors}}{\text{Mean absolute deviation}}$$

If the forecasting model used is the correct one, then the sum of forecasting errors will tend to zero. The tracking signal can have positive or negative values indicating the direction in which the model is biased. The calculated tracking signal can be compared to the pre-set values and if the absolute value of tracking signal exceeds a pre-defined value then an exception report might be generated to indicate the unsuitability of the model. Values of tracking signal above 3 show that the model may not be suitable. Alternatively, it might be desirable to keep track of how often the absolute value of the tracking signal exceeds a pre-set value, generally much lower than the value at which model becomes suspect. As the limit used in tests for tracking signal values gets smaller the probability that random occurrences and fluctuations will result in the actual tracking signal value exceeding the limit value increases. Therefore, to reduce the amount of work that must be carried out to determine the source of error, it is desirable that the tracking signal limit value should be increased. Using an Alpha factor value of 0.1, and a tracking signal value of 0.84, there is only a 1% probability that the tracking signal limit value of 0.84 will be randomly exceeded. But if a lower limit value of 0.62 is used then the probability of the tracking signal value randomly exceeding this limit increases to 5%. In such cases it is desirable to keep a count of the number of occasions on which the tracking signal value exceeds the pre-set limit. The exception report generated, shows whether there is a persistent error in the model and work should be done to analyse the possible sources of error.

The Role of the Computer in Forecasting Applications

The computer has a significant role to play in making forecasts in situations in which the number of items and factors involved is large, and where it is difficult to maintain close liaison with customers. Also in terms of stock usage and for preparing production plans for a large number of items, even simple computer-based statistical forecasting techniques are better than intuitive guesses. But where only a small number of items is involved and it is possible to have close contact with the customers, then the judgment of the sales and marketing personnel might be better.

Essentially the role of the computer in a forecasting system is to store the historical data relating to the level of demand for items to be forecast, and help the user analyse this data to detect the presence of any trends or seasonal patterns by performing, at a very fast rate, the large number of calculations involved. In the case of an equivalent manual system the performance of the required calculations is a slow and laborious process.

Most simple computer models react slowly to changes in the level of demand. For example, in an exponential smoothing model using an Alpha factor of 0.1, any abrupt changes in demand are not reflected quickly in the forecast for future periods. If, however, this change in demand is not random and reflects a permanent feature, then using the computer and adaptive smoothing techniques the value of α can be adjusted quickly to improve the accuracy of forecasts. With a manual system there is a natural tendency to over-react to forecasting errors. Very often if the previous forecast was too low compared with the actual demand, then the forecast for the next period is likely to be too high.

If the forecasting is carried out in an interactive manner, then the planner is able to adjust the forecasts immediately after they have been produced. The adjusted forecast can be used automatically as the input to requirements planning and inventory control systems. If, however, the forecasting is carried out in batch mode then it might be necessary to adopt a procedure in which the computer generated forecast is automatically input to the requirements planning module unless adjusted by means of another input to the system. A better approach is to require a manual confirmation of the computer forecast. While this procedure can result in delays it also ensures that at a later stage the managers cannot blame the computer for their mistakes.

Where the computer is used in this particular fashion, it is possible to take advantage of the computer's ability to analyse past historical data extensively and then modify the forecast in line with the judgment of the experts. The experts can take into account the additional information relating to, for example, an upturn or downturn in the economy, increased competition etc. It is difficult to build computer forecasting models which embrace all the relevant factors.

A computer-based statistical forecasting system should be able to carry out the following functions:

1 Collection and updating of historical data relating to the stock usage or product demand.
2 Detection of the existence of any trends in the historical data. The data can either vary around an average value, or have a sloping trend or finally they can exhibit a seasonal pattern.
3 Help the model builder in the development and selection of a suitable model by using regression analysis techniques.
4 Make estimates of the future demand for the product, based on the historical data and the selected model.
5 Finally, the system should have facilities to keep a check on the accuracy of the model selected for forecasting.

Fig. 6.1 shows the role of the computer and forecaster in a computer based forecasting system.

Considerations in Computer-Based Forecasting Systems

A system in which statistical forecasts are complemented by subjective forecasting is inherently better than forecasting systems based on the use of any one of the individual methods. Better statistical forecasting techniques result in lower errors, and the buffer stock required to provide a given level of service is reduced. However, the application of more sophisticated techniques, such as Box–Jenkins modelling, is expensive and time consuming.

It is usually necessary to identify first the external factors which affect the level of sales for the particular product, and then carry out extensive model building in an iterative manner. Also the forecasts based on these models may not be a great deal more accurate than those based on other simple forecasting techniques which only use historical product demand data. This is because, apart from major identifiable factors, there is usually a number of other minor factors, for example the interest rates and market conditions in the trading partner countries, which affect the level of sales. The historical usage data cumulatively reflect the effect of all these variables and although the forecasts based on historical data will not, for example, reflect the effect of a quick upturn in the economy, in practice these forecasts can be adjusted to take account of other available information.

The people using the forecasts made by these techniques may not fully appreciate the assumptions behind the techniques. Simple forecasting techniques, such as exponential smoothing, are better understood by users and are easy to apply. It is often unnecessary to apply sophisticated forecasting techniques to all the items in a company's inventory. Effort which is concentrated on the comparatively small percentage of items responsible for a high proportion of the company's turnover often leads to better results.

An accurate record of historical demand data is a prerequisite for the application of statistical forecasting techniques. The demand in the present context refers to the total demand which could be satisfied if enough stock existed, and not the demand that has been satisfied in the form of actual shipments to customers. If such accurate data does not exist or cannot be collected for any reason then it will be necessary to use subjective forecasts. Alternatively, if the demand data for a particular period is missing then an average value for the missing period can be substituted.

The selection of the forecasting time interval is an important factor in the development of a computer-based forecasting system. The time interval should be neither too short nor too long, and should be chosen to reflect the actual fluctuations in demand and not just a levelled demand. Short term forecasting techniques are not really suitable for dealing with data which move infrequently. The time interval selected should be such that some data movement takes place during the sampling interval. However it must not be made so large that the system fails to respond quickly to genuine changes in the demand pattern.

The demand data for the current period should be carefully vetted to test its validity, and smoothed to remove the effect of unusual events, for example, a large sale before an increased tax comes into effect or a large export order unlikely to be repeated. A large fluctuation can be detected by building a demand filter which checks that the demand lies between pre-set limits, for example, current average value ± 5 MAD. If the demand does not lie within these limits then an exception report can be produced, and the forecaster asked to confirm the demand manually. Otherwise an average value can be used as the demand for the period.

The amount of historical data required for the application of forecasting techniques varies from one item to another depending upon the model to be used. A minimum of twelve points should be used for important items without any seasonal demand pattern, in which case a minimum of two years demand data are required. More data points show better results. The data should refer to demand for fixed length time intervals. For example, four week time intervals should be preferred to monthly time intervals.

Almost all types of demand data falls into one of the following categories:

(a) A constant model.
(b) A trend model.
(c) A cyclical or seasonal model.
(d) A model which exhibits a combination of patterns described in (a), (b), and (c).

Therefore a comprehensive forecasting system should provide for models which can be used to deal with all these different types of demand patterns. The model selection module should be used to test the time-series for the existence of any trends or cyclical variations. A graphical plot of the historical time-series values is particularly useful at this stage. Ideally this particular time-series analysis and model selection should be carried out in an interactive manner. Regression analysis techniques can be used to establish the basic type of model which should be used for a given set of data. The sum of squares of the errors encountered, when different models are being used, can help in selecting the best model. Once the appropriate model has been selected the values of various factors, i.e. the various smoothing constants, trend factor, seasonal factors, etc., can be determined and stored on the part master file along with other required data items such as the current smoothed value.

The single exponential smoothing constant or the Alpha factor can have a value between zero and one. In practice its value is usually chosen to lie between 0.1 and 0.2. The lower value of Alpha results in a slow response to changes in the level of demand. However, too high a value can result in the system responding too quickly to random changes in demand pattern. The value of Alpha can be selected on a trial and error basis. If enough historical demand data are available then it is possible and desirable to carry out a simulation study using different values of Alpha, and the value which results in the smallest sum of squares of forecasting errors is selected.

Similarly optimal values of the Beta (β) and Gamma (γ) factors used to update the trend and seasonal factors respectively can be determined using simulation studies or on a trial and error basis. It is suggested, on the basis of past experience and simulation studies, that the initial values of the Beta and Gamma factors should be selected as half and twice the selected value of the Alpha factor. Similarly it is suggested that the Delta factor (δ), used to update the value of mean absolute deviation, should initially have the same value as the Alpha factor. In some cases, for example when a new product is introduced and no historical demand data exists, it is not necessary to use the model selection module. In the absence of enough historical data an arbitrary double exponential smoothing model, based on past experience, can be selected initially provided the product demand does not exhibit cyclical or seasonal variations. However, the system should provide an option whereby the users can decide on the type of model and the values of parameters to be used for forecasting purposes. Once the suitable model and associated constants have been selected the forecasts can be generated easily using exponential smoothing techniques.

Following the selection of the appropriate model and smoothing constants it is necessary to calculate the initial values of smoothed averages. In a flexible system these might be calculated automatically by the system or specified by the user. The second procedure will have to be used if enough historical data are not available for calculating initial values, when the average can be estimated and the trend value set to zero.

The use of exponential smoothing techniques, in the absence of enough historical data, can result in the calculation of biased forecasts. Therefore the system should provide a feature whereby forecasts are based on the use of moving average values of the demand. It is desirable to continue this procedure until demand data for 1/Alpha (α) time intervals become available.

The application of exponential smoothing techniques does not require storage of historical data, although the management might wish to store it for analysis purposes. It is only necessary to retain current values of parameters such as smoothed average, smoothing constant, trend factor, seasonal factors, mean absolute deviations. As new demand data become available they are used to update the values of all the relevant parameters. The computer can be used to calculate the forecast for as many future periods as considered necessary. The time horizon for which forecasts are made is dependent upon factors such as the product lead time, and the purpose for which the forecast is made, i.e. whether it is used for short-term or long-term planning. The length of forecasting time periods can also be varied. For example, for the first few weeks forecasts can be made at weekly intervals, followed by four week and thirteen week intervals. The longer the period over which forecasts are made, the higher the possibility of error. Hence, it is essential to review carefully the long term forecasts. The forecasts should be revised as soon as new data become available. New forecasts for important items can be prepared at weekly intervals, and the forecasts for minor items can be made at 4 week and 13 week intervals.

Since most of the mathematical forecasting techniques involve the use of historical data, it is difficult to take account of a sudden increase or drop in the level of orders. Therefore for the development of an efficient forecasting system it is essential that forecasts are modified, by experienced personnel, in the light of all the other external factors which cannot or have not been taken into account in the development of the forecasting model. This modification of forecasts is particularly important in the case of finished products, i.e. products such as assemblies and spare parts with an independent level of demand, since the demand for lower level items is dependent on these forecasts. Based on the demand or sales forecasts for the finished product the demand for individual components can be calculated using the Bill of Materials Processing (BOMP) programs and requirements planning techniques. If any attempt is made to forecast the demand for dependent items, irrespective of the demand for higher level items, then their stock level may not satisfy the assembly requirements. Even assuming that it is possible to have a very high probability (95%) of ensuring the availability of individual components, the probability that 12 components required to assemble the finished product will be available is reduced to 54%. Under such circumstances the management has to resort to splitting and expediting production orders so that the components are available for assembly. It is necessary to use material requirements planning techniques for determining the demand for dependent items.

Many items in stock exhibit a lumpy demand pattern, i.e. in some periods there is a high demand followed by a very low, often nil, demand in other periods. Normal regression analysis techniques cannot be used to select a model for such items. In principle this difficulty can be overcome using group forecasting techniques in which a number of similar items are grouped together. The percentage contributed by each individual item to the overall volume is determined. The model and associated parameters are selected using the normal procedures, and an overall group forecast is prepared. The product of group forecast and the percentage contribution of an individual item to the overall volume, gives the forecast for the item.

The development and maintenance of an appropriate forecasting model is a continuous process. The major effort is required when the model is first built. However it might be necessary to make considerable modifications to the model at a later stage. For example it will be necessary to modify a model developed during a period of relatively stable demand to reflect changed conditions during a period of declining product demand.

Forecasting Module

The input to the forecasting module will be the actual demand in the current time period. This latest demand data should first be vetted, using the appropriate demand filter, and if the demand does not lie within specified values then an exception report should be generated and the users asked to either confirm the demand or supply a smoothed or average value. The system should also incorporate a facility to compute an average value automatically. This will be followed by the calculation of the error between the forecast for the current period and the actual demand. The model to be used can be specified by a variable parameter and, based on the values of the various smoothing constants, the forecasts for future periods can be calculated. The number of future periods for which forecasts are made should be flexible and user specified by means of parameters. Apart from a printed report showing the forecasts of future demand, a record should be kept on the part master file so that the forecasts can be input automatically to the master production schedule system. A record of the value of the sum of squares of the errors should also be kept and if this value exceeds a pre-set value then an exception report should be produced so that an analysis of the possible sources of error can be carried out.

The format of the forecasting report should be flexible, and it should be possible to include in the report items such as: forecasts for future periods; Alpha, Beta, Gamma, Delta, and seasonal factors; mean absolute deviation; sum of the squares of errors; value of tracking signal. If any unusual conditions are encountered during the processing of data an exception report which highlights the condition should also be generated.

Summary

The management should not over-rely on the forecasts prepared using mathematical models; the model and the resulting forecast is only as good as the data used to build the model. All statistical forecasts are based on the assumption that the past conditions and behaviour will continue to be reflected in the future, which may or may not be true. It is impossible to build an all embracing model which can take account of all the possibilities. It is desirable, and indeed necessary, to modify these statistical forecasts in the light of additional knowledge of the market place.

7
Order Processing

Introduction

The survival of a manufacturing company in the fiercely competitive international business environment depends upon a number of factors, such as the quality and prices of their products, the delivery lead time, and the manner in which the orders are handled. If it takes a long time for the sales department to provide a quote in response to customer enquiries, then the customer might decide to place his order elsewhere. Similarly following the placement of an order, if the sales department is unable to answer queries relating to the current position of the order, then the customer might take his future custom elsewhere.

The order processing system is used to handle customer orders from the time of receipt until they are shipped, and the customers are invoiced. The other main functions of an order processing system are to handle customer enquiries and provide price quotations.

To ensure that the customer enquiries are efficiently and quickly handled, the sales personnel must be provided with suitable tools. An on-line real-time computer system is an ideal tool for performing the order processing functions. The order processing system and the sales personnel form the interface between the customer and the manufacturing departments of the company. Since the order processing system is the only company function with which outside customers come into direct contact, it is desirable that the other company systems should be carefully dovetailed into it. The order processing data can be used as the basic input to the commercial and accounting systems used by the company. Many companies often neglect this aspect of their operations. Each order received by a manufacturing company has implications which vary according to the current activity state.

The processing of sales orders makes an impact on almost all the activities which take place in a manufacturing company. The availability of items required to satisfy the customer order is determined by obtaining access to the inventory records. The actual customer orders are subtracted from the forecast of the number of items required during a particular time period. These customer orders are also used as an input to the master production schedule. The discrepancies between actual orders and forecast data are used in the calculation of safety buffer stocks. Material requirements planning, capacity planning, shop load levelling, order release, and operations scheduling activities are based on the master production schedule and any changes which might be made to it. When customer orders are changed the effect of these changes on the production schedule, and the following activities has to be determined. The current position of production orders on the shop-floor is monitored continuously to ensure that they are completed on time, and that the delivery dates are satisfied. Following the assembly and inspection of the product, it is dispatched to the customer along with the invoice. Once payment is received, the order can be deleted from the current records. The ability of the sales department to carry out the various order processing functions efficiently is dependent on the availability of accurate and up to date information. With manual systems where a large volume of paperwork has to be handled, the information is not always accurate and may not be available on time, often resulting in increased delivery periods.

Computer-Based Order Processing Systems

A computer-based customer order processing system makes a considerable contribution to the efficient operation and profitability of a manufacturing company. The time gap between the receipt of a customer order and the start of production work on it can be minimized. The order can be input to the system as soon as it is received. All the necessary checks on the order can be performed automatically as part of the order processing system. The amount of clerical effort and time required to process the order is substantially reduced. This saved time can be used to improve the competitive position of the company by quoting shorter delivery times. It also enables the management to carry out more detailed production planning. The load on manufacturing facilities can be balanced. As the amount of detailed manual effort required to handle the order is reduced, the possibility of the order being misplaced, lost or delayed is minimized. Depending upon the numbers of orders received, one or two sales office clerks can enter these orders to the computer system and from then onwards a permanent record is maintained on the computer files until the time when the order is deleted.

In an on-line system terminals can be used to enter orders. The operators can be guided through the whole of the order entry process. Pre-formatted screens minimize the time and effort required to enter the orders. The possibility of errors creeping into the system is also reduced since the amount of detailed data which must be input to the system is relatively small. For regular customers, the standard data such as customer name, delivery and invoicing addresses, can be retrieved automatically from the database, and there is no need to key-in this data again. By keeping track of the previous orders, the system program can automatically generate the next order number. The system programs can be designed to check that all the required data are input to the system. If the data are inadequate or inaccurate for a complete description of the order, then the system can reject the order until such time as complete and accurate data are input, when order records can be created. These records can then be used to prepare all the required documents such as order acknowledgement, invoices, packing lists, labels, in addition to forming basic input to other computerized systems used for production planning and control, commercial sales analysis and accounting.

As soon as the order is ready for shipping to the customer, an invoice can be generated automatically by the order processing system and despatched with the order. The shorter invoicing time improves the cash flow situation of the company.

Credit check can be performed at the time of order entry. The order can be rejected if the acceptance of this order will result in the credit exceeding a pre-set limit. This can then be brought to the attention of the senior management personnel who may, if necessary, over-ride this credit limit and the order can then be processed in the normal way. As a result the losses incurred due to bad debts are reduced.

The sales department personnel can check, via a terminal, the possibility of satisfying the order ex-stock. A frequently encountered problem of the same stock being inadvertently promised to different customers by two sales clerks and also scheduled for use in the machine building program by the production personnel, can be overcome. The sales clerk can use a terminal to enquire about the available free stock and, if necessary, allocate it to a particular customer. The part master file can be updated immediately to reflect the effect of this transaction. If such free stock is not available, then the customer can be quoted a realistic delivery date by taking into account the released and planned orders. This process may be taken one stage further for large orders. The effect of a large order on the available capacity can be simulated before quoting a delivery date.

The interface between the sales order processing and work-in-process control systems can be used to handle customer order enquiries efficiently as the order progresses on the shop-floor. The terminals can be used to make these enquiries. If the order is not making adequate progress, then an exception report can be produced to highlight this condition and bring it to the attention of the management.

Once the order is delivered, the management can use the system to keep control over any delays in payment of the invoice, thereby contributing to the improved cash flow situation of the company.

Features of an Order Processing System

A computerized order processing system should be capable of carrying out the following functions:

1   Order entry.
2   Handling enquiries relating to orders being processed.
3   Controlling orders as they make progress on the shop-floor.
4   Providing quotations in response to written or telephoned customer enquiries.
5   Production of shipping and invoicing documents.

These functions are now considered in more detail.

Order entry transactions can relate to new orders received from the customers, or amendments to existing orders. When a new order is received it should be edited before creating a record in the customer order file. The first check should show whether it is an existing customer or a new customer. In the case of new customers, it would be necessary to create the standard customer data such as customer code number, customer name, customer address, credit limit, any discount code which should be used in invoicing calculations etc. In the case of old customers, the standard data such as customer address, and shipping address etc., can then be displayed on the screen and compared against the information contained in the order. Any necessary modifications to the standard data, such as change of customer address, can then be made. At this stage the variable data such as items required, their quantities, the delivery dates, the temporary shipping address, etc., can be input to the system. The system should then check, for all orders, the credit limit and whether it would be exceeded if the order is completed.

If the credit limit is exceeded, the order should be rejected and an exception report produced for management attention. The credit controller can then contact the customer and tell them that they should make some payments before the processing of further orders can be continued. He might be aware of some payments which are in the post, or some verbal agreements to raise the credit limit in the event of extenuating circumstances, when he can override the computer's rejection of the order. The system programs should include this over-riding facility. However, only senior management should be able to exercise this privilege. Once all the required data has been input to the system, an internal order number can be automatically allocated to the customer. An order acknowledgement form can then be produced automatically by the system. This should include the internal order identification number which the customer should be requested to quote when making enquiries about the status of the order. Also, at this stage, a higher external priority can be allocated to urgent orders. The assignment of high external priorities should be carefully controlled, otherwise the priority system will break down and lead to severe problems. The next step would be to check whether the item required can be supplied from available free stock. If this is the case then the available free stock can be allocated to the customer, and the necessary shipping and invoicing documents can be prepared. Considerable savings in the disc memory space required can be made by linking the variable order data to the standard data for the customer. All the orders for a particular customer should be chained together. Similarly all the separate items within an order should also be chained together. This chaining procedure is as shown in Fig. 7.1.

This linking of records enables the users to retrieve information relating to all the orders for a particular customer quickly, without having to search through the whole of the database. A printed out report of the up to date status of all the active orders for a customer can be very useful to a salesman when he visits the customer.

Amendments to Existing Orders

It is sometimes necessary to make amendments to orders already input to the system and currently being produced on the shop-floor. The customers often telephone to enquire if it is possible to provide an early delivery, delay the order, alter the order quantity, or cancel it altogether. Many companies have standard order cancellation charges. In other cases it might be necessary to estimate the costs incurred so far in producing the order, and the management can then decide the total cancellation charges. A computer-based order processing system and its interface with the production systems can be used to examine the various possibilities before giving a reply to the customer.

The total costs incurred on an order may be determined by retrieving the up to date record of material, labour, and overhead costs. These costs can be accumulated and the customer informed about the cancellation charges. If the customer still wishes to cancel the order, then it can be deleted. The management

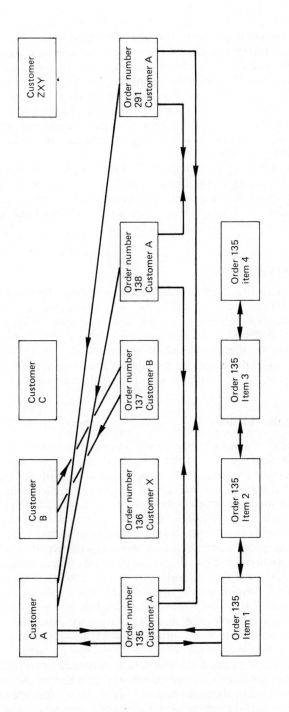

Fig. 7.1 Chaining of all the orders for a particular customer, and chaining of items within an order

can decide whether to continue with the production of the whole order or individual components and use the items for ex-stock deliveries.

The possibility of early delivery can be examined by reviewing the total volume of work-in-process, and the percentage of this work which has been accorded a high priority. If only a small volume of work has high priority, and if adequate capacity exists, then it would be possible to make an early delivery. The new delivery date can be simulated using the capacity planning and shop load levelling procedures, further discussed in the following chapters, and a new delivery date quoted to the customers. The impact on the remainder of the work-in-process of allocating a high priority to a particular order, can also be examined using these simulation procedures.

The possibility of delivering an increased order quantity can be checked by retrieving the records relating to the available free stock, released production orders, and planned orders. If unallocated stock does not exist then the effect of increased order quantity on lower level items can be determined, and, if necessary, a new delivery date quoted to the customer.

Handling Customer Enquiries

The customer enquiries relating to orders currently being processed can be easily handled in an on-line real-time order processing system. When an enquiry is received the information relating to the relevant order can be retrieved by entering the internal order identification number. If the order identification number is not available, then it would be necessary to retrieve summary information relating to all the orders for the particular customer, and the up to data status of the relevant order number can then be determined.

Control of Customer Orders

The main objective of a customer order control system is to ensure that orders are delivered on the date specified in the acknowledgement form sent to the customer. It is necessary to monitor the progress of these orders at regular intervals to ensure that they are not held up too long in any particular area, and that they are making adequate progress. It might be the management policy to peg all production orders to individual customer orders, in which case their progress can be followed easily. It is possible to develop, on the basis of historical data, the standard times which an item should spend in a particular area. These times or decision rules can be built into the system programs. The activities on the shop-floor can be monitored continuously and if it is found that the order is spending more than the allowed time at any particular stage, then an exception report can be generated. More often, however, individual customer orders are consolidated into one production order, in an attempt to reduce manufacturing costs, and it is difficult to follow the progress of an individual order. In a computerized work-in-process control system the programs can be used to keep track of the progress made by production orders; if particular operations have not been completed by specified dates an exception report can be produced to bring this to the attention of the management. Also if the orders are late the internal order priority, used during the operations scheduling process, increases and the order will appear high up in the work sequence lists for the following day/shift. In addition the management can, if necessary, allocate a higher external priority to the particular production order. The work-in-process control, and operations scheduling systems are discussed in detail in the following chapters. Once the order is dispatched to the customer, the order records can be deleted or, if required, input to the commercial and accounting systems.

Delivery Date and Price Quotations for New Orders

An on-line computer system can provide considerable assistance in replying to enquiries with regard to prices and delivery dates for new orders. The order clerk can determine the price and item availability by simply entering the item identification number and retrieving the required information from the part master file. This information can be printed out and sent to the customer.

Customers often make enquiries about the availability of items, particularly spare parts, before placing a definite order. It may be some time before this order reaches the supplier; in the meantime the same item may be supplied to another customer from whom an order has been received. Alternatively this item may have been used up in the machine building program. Some customers may be willing to accept such risks. In general, however, this can create a bad feeling between the customer and the supplier. In order to improve customer service levels, it is necessary to safeguard against such happenings.

One possible solution is to create 'special' orders. The required quantity can be allocated to the special order, and the customer informed that the item will be delivered if a definite order is received within a specified number of days. In the meantime this 'special' order quantity cannot be allocated to other orders. If the order confirmation is not received within the specified number of days, then this 'special' order can be automatically cancelled, and the released quantity becomes available for general use.

Production of Shipping and Invoicing Documents

After the production of the required order items is complete, or a few days before the completion date, the order processing system can be used to prepare the shipping documents and the invoice. The customer and order records can be retrieved by means of appropriate programs, and the pick up lists containing details of all the order items, the packing instructions, and the labels can be produced automatically. The invoice can be prepared by obtaining access to the relevant information in the customer, order, and part master records. After the order is despatched and the payment received from the customer, the order can be deleted from the current files, and, if necessary, summary information can be retained for historical sales analysis purposes. In addition the relevant data can also be passed on to the accounting systems. There is often a statutory requirement to keep details of all financial transactions for a specified period of time. Such information can be stored on microfilm or other suitable storage media.

E

Order Processing Programs

The flow of information in a typical order processing system is illustrated in Fig. 7.2. Detailed data for individual items can be retrieved from the part master file. The standard information relating to customers is stored in the customer file. Pointers are used to link the standard customer record to the variable information for individual orders, included in the open order file.

Typical contents of the record for a particular customer include the identification number, name, head office address, shipping address if different from the head office address, invoicing address if different from the previous two addresses, name and telephone number of the person who can be contacted for enquiries relating to the orders received from this customer, credit limit, outstanding customer credit, customer discount code, usual shipping method, shipping instructions. Similarly the fields typically contained in an open order file include customer order number, customer identification number, date on which order was received, part number of the item for which an order has been placed, order quantity, order due date, shipping address if different from the one in the customer file, status code which indicates the current position of the order, shipping instructions, amount of invoice for this order. In addition the manufacturing companies will have their own parochial requirements which can be satisfied by including relevant fields in the database.

Within the open order file all the orders for a particular customer can be linked together using the forward and backward chains. In addition all the items included in one order can also be chained together.

The number of programs included in an order processing system, and the complexity of individual programs will vary according to the number of functions performed as part of the system. The data processing necessary to carry

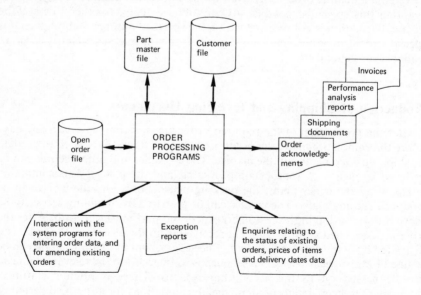

Fig. 7.2 Flow of information in a typical order processing system

out any particular function has already been discussed in broad terms. The detailed system specifications will differ from one company to another. However, the system should include facilities to validate input data, check all orders for credit limits, create a record in the customer file for new customers, create open order file records for all new orders received and accepted by the system, amend existing open order file records, and produce shipping and invoicing documents. No attempt should be made to produce one large program which can carry out all these tasks. Modules can be written and debugged individually before running the complete system program. These program modules can be called into the main memory as required.

The output from the system programs should include order acknowledgements, shipping documents and invoices, performance analysis reports, and any exception reports highlighting conditions which should be brought to the attention of the management. The number of routine reports, and the details included in them, should be kept to a minimum.

Summary

A computer-based on-line order processing system enables the management to improve customer service levels by reducing the order processing and total delivery times, and also the time required to react to changes. Customer enquiries can be handled easily and quickly. The progress of the orders, from the time of their input to the system until their shipping, can be easily monitored and controlled, thereby reducing the number of late orders. Losses due to bad debts are reduced by checking the customer credit limit before accepting an order. The manual clerical effort required to process the orders is vastly reduced. A small amount of manually input data can be used to produce all the required documents such as order acknowledgement form, shipping papers, and invoices. Conditions requiring management attention can be highlighted by exception reports, thereby reducing the need for detailed minute-to-minute intervention.

8

Creation and Amendment of Master Production Schedule

Introduction

The master production schedule, also referred to as the assembly build programmme, represents the requirements for end items over a long period of time. It shows the quantities of individual items and the dates by which these quantities should be manufactured. The estimates of future demand prepared by the management or produced using the statistical forecasting system and the actual customer orders form the input to the master production schedule. As a result the schedule is very dynamic and in a continous state of flux. As forecasts are updated to take account of the latest demand data, the master production schedule also has to be revised. While the computer can be used to handle all the details, it is nevertheless necessary to use skilled human judgements in the preparation of this schedule. The current customer order book may not be full and the forecasts of future demand may be low during periods of recession. The statistical forecasts represent the future trend based on recent past demand data, and are valid only if the current conditions do not change. Under these circumstances the management will have to decide whether to utilize the available machine and manpower capacity, and manufacture for stock in the hope of an upturn in economic activity. Also the delivery lead time is rarely adequate to start production from scratch. As a result it is often necessary to manufacture common sub-assemblies and components before the receipt of the customer order. Once again the management has to use its intuition about the future demand for its products.

In the preparation of the master production schedule, it is also necessary to take into account the availability of materials, machine and manpower capacity. A large number of combinations can be used to manufacture all the items required over a given time period. The criteria used to arrive at a production schedule vary according to the current conditions. During periods of high activity the company will wish to use a production schedule which maximizes its profitability, whereas during low activity periods it might be necessary to use a schedule which, while covering overheads, minimizes the need for short time working or redundancies. It is unrealistic to attempt to experiment with different production schedules on the shop-floor. The need for such shop-floor experimentation is avoided by using the computer simulation facility. The impact of various production schedules on material, machine and manpower requirements can be examined in the simulation mode before finalizing the production schedule for a particular period. Similarly as a revised forecast is made, the implications of the change can quickly be simulated. These simulations are only possible due to the availability of a computer which can carry out all the required calculations at very high speeds. With manual systems it is difficult to determine the full impact of alternative production schedules and any changes made to them. In a computer-based system the production planner and the computer can become equal partners in the preparation of the short term production schedule. In the short term the requirements are reasonably firm while the long term requirements are rather uncertain and subject to frequent revisions. It is often necessary, even in the short term, to add urgent customer orders to the production program. Rework items and additional requirements due to excessive scrap on the shop-floor also have to be included in the production schedule.

In general the production schedule shows the requirements for end items. In some cases, however, it might be necessary to include the requirements for major assemblies, for instance in the cases of aero-engines or power generation equipment which can be divided into a number of major assemblies.

In the short term, the master production schedule can be used for material requirements, capacity planning and shop load levelling calculations. In the long term this schedule is useful for estimating the long term capacity requirements and the capital required to provide any additional capacity. The purchase and commissioning of new machinery often takes a long time. Similarly, skilled manpower cannot always be recruited at very short notice, nor is it practical, in the interests of good industrial relations, to make employees redundant every time there is a slack in the economic activity.

In view of the fact that the master production schedule shows the quantities required and the associated due dates for a large number of items, it can also be used to allocate priorities to different orders. It is, however, necessary to ensure that the master production schedule is up to date and realistic, i.e. full account should be taken of the total capacity available in any individual period. The production schedule should be adjusted, if necessary, so that it takes account of the shop-floor conditions. An unrealistic production schedule will only result in a situation in which the priorities allocated to production orders cannot be adhered to on the shop-floor.

It is important to appreciate the distinction between the forecasts of future demand and the master production schedule. The quantities and dates, included in this schedule, represent actual commitments resulting from:

(i) Customer orders received.
(ii) Forecasts, prepared by the sales and marketing personnel or using statistical routines, which have been subjected to value judgments.

Features of a Master Production Schedule Creation System

The master production schedule forms the input to the requirements planning module which explodes these assembly and end item requirements into lower level sub-assemblies, components, and materials needed to fabricate the final product. All the future planning and actual work on the shop-floor depends upon the contents of this schedule. In order to ensure that the schedule is realistic and takes account of the necessary constraints, the data processing system used to prepare it should include the following facilities.

There should be facility to maintain a master schedule for all the items, including spare parts, which can be ordered directly by a customer, as well as all the large assemblies into which a product is split. It should provide information which can be used by the management for making decisions about the best production schedule for individual items and the total number of items required in a particular time period. The system should have the capability to make estimates of the long term machine capacity, manpower capacity and capital requirements. Also it should be possible to use the system for determining the gross and net capacities, of individual departments, required to satisfy a given production schedule. It should not be necessary to re-process all the data every time a change is made to the master production schedule. It should be possible to calculate simply the effect of the changes that have taken place. The system should have a simulation facility so that the effect of alternative production schedules can be assessed before making decisions. The conditions which require management attention or intervention should be highlighted by means of exception reports.

Creation of the Master Production Schedule

The basic input to the schedule consists of forecasts of future demand and customer orders. Forecasting is suitable for companies engaged in repetitive manufacturing, for example for consumer products. However, statistical forecasting is unreliable in the case of custom built products, such as aero-engines, oil tankers, or power generation equipment. In these cases it is better to rely on the estimates made by sales and marketing personnel who are in close contact with the customer and production is not started until customer orders are actually received. The majority of manufacturing companies lie somewhere in between these two extremes and it is necessary to consider forecasts as well as customer orders in the development of the schedule.

The statistical techniques, used for making forecasts of future demand, have been discussed in considerable detail in an earlier chapter. The definite customer orders can be allocated against the forecasts for the period in which the ordered items have to be delivered. If, however, the total number of orders exceeds the forecast, then it would be necessary to check the possibility of including the order in the master schedule for the period under consideration. This can be carried out by examining the availability of materials/components as well as the capacity required to fabricate the final product. This task can be eased by keeping a record of the total machine/manpower resources required to manufacture each end item. In the event of non-availability of required materials or capacity, a realistic delivery date can be quoted to the customer. It is often necessary to accept urgent orders from long standing customers even though the required capacity may not be available. In such cases the requirements for overtime working and the resulting increased labour and expediting costs can be simulated and included in the price quoted to the customer who can then decide whether or not to place the order.

As definite customer orders are received, the computer can be used to peg the scheduled production quantity to individual orders so that the same quantity is not inadvertently allocated to two different customers.

The production schedule for an item, in a given period, can be determined by means of the following relationships:

Requirement in a given period = Forecast of future demand (or Actual customer orders) + Safety stock

Production schedule for the period = Requirement for the period − Available free stock

Amendments to the Master Production Schedule

The production schedule has to be updated continuously in order to ensure that current conditions are fully taken into account. The amendments to the schedule can be due to a number of reasons. For example, the forecasts may have been too pessimistic. Alternatively a key machine might break down for a relatively long period of time resulting in a hold up in production work. Also it may not be possible to work the required amount of overtime due to the current industrial relations atmosphere. If the schedule is not updated to reflect the current situation, then the priority system used to expedite urgent orders will quickly break down. The management can use the simulation facility to decide which orders should be rescheduled and then prepare a new and realistic production schedule. The simulation results should be stored in a different area of the disc so that they do not automatically update the affected files. Once the management has decided on the best production schedule, then relevant records are modified to take account of the necessary changes. The application of on-line real-time systems is particularly valuable for making such policy decisions. In a batch system the simulation results may not become available until the following day, whereas the necessary decisions have to be made within the next half an hour or so. In addition a new simulation might have to be carried out upon receipt of the previous results, in which case it will be necessary to wait for another day. Such delays often mean that the simulation facility is never used in practice. In an on-line real-time situation the effect of alternative policy decisions can be examined quickly, in an interactive fashion, before arriving at the final decision. Similarly the management can use this simulation tool for examining the effect of a changed product mix. The amount of data processing required to simulate a given situation is substantially reduced by using a system which simply calculates the effect of changes to a previous situation. The same facility is also useful for actual updating of files. The determination of the impact of changes on the resources required is simplified by storing, on computer files, summary information relating to the resources required to produce each end item.

Resource Requirements for End Items

An estimate of the resources required to manufacture one unit of each end item can be computed by exploding its product structure which shows the components required and the associated quantities. The parts explosion process was discussed in the Bill of Materials Processing (BOMP) chapter. These BOMP programs can be used in conjunction with the part master file, the product structure file, and the routing file, for calculating the load on individual resources and the time at which this resource is required for producing the end item on the due date. The time required to produce one unit of an item is estimated by adding the run time to a specified proportion of the set up time. Although this rough and ready estimate of the resource requirements may contain errors, it does allow the management to estimate quickly the effect of changes to the production schedule without the necessity to process a large volume of data. This facility can be very useful provided that the errors due to the difference between the actual and estimated resource requirements are small, and acceptable to the management. The information relating to the load and timing of resources required to produce a unit of a given end product can be stored on the End Item Resource Requirements File. A unique code can be allocated to each resource referred to in this file which should be linked with the part master file by means of suitable pointers. The use of BOMP programs for calculating the resources required to produce end items is illustrated in Fig. 8.1.

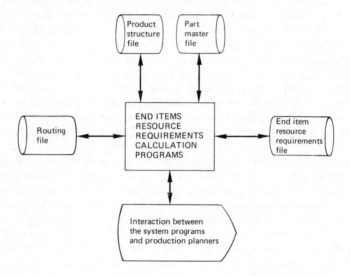

Fig. 8.1 System programs required for calculating resource requirements for end items

Master Production Schedule Creation and Amendment Programs

Fig. 8.2 shows the flow of information in an on-line real-time system used for planning and updating the master production schedule.

The forecasts of future demand, prepared by the statistical forecasting system, can be automatically input to the system following modifications which the management might wish to make to take account of any external factors. This information can be stored in appropriate fields on the part master file or its extension. The revised forecasts, as they are produced, update the previous data. The definite customer order requirements, input by the open order file records, can be checked against the production planned for the period. If the required order quantity can be satisfied from the available production planned for the appropriate period, then the customer order can be pegged with the relevant planned production order. If, however, the order cannot be satisfied by the planned production order quantity, then the possibility of revising the

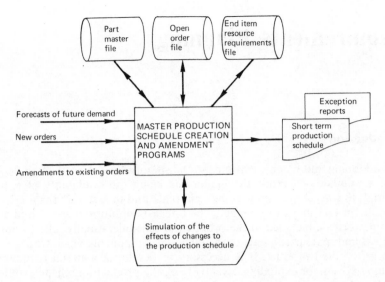

Fig. 8.2 Flow of information in an on-line real-time system used for the creation and updating of the master production schedule

production schedule is examined. At this stage the End Item Resource Requirements File can be used to assess quickly the effect of change to the production schedule for the required item. The resulting overload on some of the resources can be calculated and printed out in the form of an exception report. If the resulting overload is not acceptable then the order may be allocated against the planned production for the following period, and a realistic delivery date quoted to the customer.

In addition to updating the relevant records, the output from the system programs should consist of the planned production schedules and the exception reports which highlight the overloaded work centres.

Summary

The activities on the shop-floor are primarily dependent on the contents of the master production schedule which forms the input to the requirements planning, capacity planning, and operations scheduling systems. This schedule should be realistic and continuously updated to reflect the current conditions in the market as well as on the shop-floor. The simulation facility, used as part of the master production schedule creation and maintenance system, enables the management to examine quickly the effect of changes to the production schedule and a different product mix. The customer orders can be allocated against the planned production thereby reducing the number of ad hoc changes necessary on the shop-floor. The effect of an urgent customer order on the remaining production schedule can also be quickly simulated. As a result the management becomes aware of the full impact of its decisions.

9
Requirements Planning

Introduction

In a jobbing and batch environment, the smooth flow of production depends upon a number of factors, the major one being the availability of required materials in correct quantities at the right place and time so that the production schedule for end items can be met. The forecasts of future demand and actual customer orders, included in the production schedule, usually refer to finished products and spare parts, and not to individual components which are assembled to form the final product. It is necessary to develop a material requirements planning system for exploding, level by level, the production schedule and then calculate the quantities and types of materials, components and sub-assemblies which must either be manufactured or purchased in different time periods. Complex products require a greater degree of requirements planning than simple products. Any system used for net requirements planning must take account of the currently held inventory as well as work-in-process. At each level the generated gross requirements have to be netted and batched according to the rules specified for individual components or raw materials.

A major objective of the requirements planning system is to exercise good control over the levels of inventory and simultaneously assure that management objectives, in terms of the required service levels, can be satisfied. Depending upon the nature of demand, two different types of techniques can be used to control the level of inventory:

1   Reorder point technique is useful for independent items, i.e. those items whose demand is unrelated to that for other products. Typical examples of independent items are finished products, spare parts, and optional features available on end items. The application of the above technique is based on an anticipation of the level of demand for these items. Forecasting techniques, discussed in chapter 6, can be used to predict the future demand levels.

2   Material requirement planning technique is used for dependent items. The demand for such items depends upon the requirements for end products and other items, used as spares, at a higher level in the product structure and can be calculated using material requirements planning techniques. Reorder point techniques should not be used for these dependent items and no attempt should be made to forecast their future demand independently. Independent demand forecasts for components, assembled to form the finished product, usually result in a mismatch of the components required or lead to much higher stock levels. Assuming that the inventory control system assures 95% availability of individual components, then the probability that 8 components required in the assembly of a particular product will be available in stock, at the required time, drops to 66%. This would be deemed to be highly unsatisfactory and to satisfy a

particular customer order it would be necessary to expedite the manufacture and availability of required components. This expediting usually results in a vicious circle. The planned orders get behind schedule resulting in an increased number of other customer orders which are not satisfied on time. It is therefore desirable that a material requirements planning system should be used for dependent items.

Another objective of a material requirements planning system is to recommend the placing of production or purchase orders so that inventories depleted, as a result of withdrawal of items, can be replenished. This entails a continuous monitoring of future requirements as well as stock levels. By considering the product lead time etc. action can be taken, at the appropriate time, to ensure adequate supply. Wherever possible advantage can be taken of lot sizing techniques involving different versions of economic order quantities.

The computer can be used to accumulate the period by period demand for individual items. Orders can be placed when the quantity on hand is inadequate to meet this accumulated demand. Fig. 9.1 shows the accumulated demand for an item in future time periods.

It can be seen that the quantity on hand covers the full demand for periods 1 and 2, and part of the demand for period 3. Therefore in order to prevent a stock-out condition, additional orders must be placed, taking into account the lead time, to cover the demand for periods 3, 4 and 5. A variety of lot sizing techniques, discussed later, can be used to calculate the order quantity.

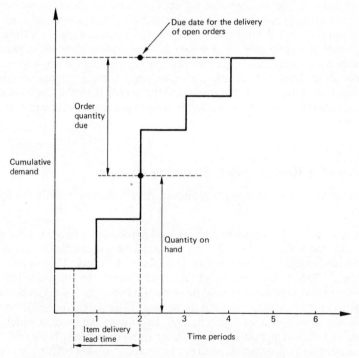

Fig. 9.1 Illustration of the accumulated demand for an item

The following discrete steps are involved in a level by level, period by period, material requirements planning system.

1   Computation of gross requirements for individual items. The input to this phase consists of production schedules for the end items, and the requirements for lower level components determined during the parts explosion process.
2   Calculation of net requirements for the required items.
3   Batching of requirements and calculation of order quantities.
4   A number of single level explosions carried out to determine the requirements for lower level components. During the parts explosion process the requirements for a particular component, generated by an item at a higher level, are combined with other requirements for the same component. The due date for these quantities is offset by the appropriate lead time to determine the release date of planned production/purchase orders.

Planned orders refer to orders which have not yet been released either to the production shops or to outside suppliers. By keeping the order release separate from the requirements planning process it is possible to review the order prior to its release. A periodic review of planned orders tells the management that an order is due for release. The order can be released either automatically within the system or manually after management review. The details of the 'order release' process are further discussed in a later chapter.

Such a requirements planning system results in a series of planned orders which are released, at the correct time, to satisfy the production schedule requirements. It is necessary to use a computer for such a material requirements planning system since manual handling will require excessive manpower and time. In fact with a manual system, it would be impossible to calculate accurately the absolute requirements at any instant of time due to the frequent changes in production plans caused by cancellations of existing orders and receipt of new orders. With a computer all the level by level, and period by period, requirements are computed during one run of the system programs and the overall planning cycle time period can be substantially reduced. It is also possible to maintain a much tighter control over items.

Features of a Requirements Planning System

A comprehensive requirements planning system should include the following features.

1   The system should carry out level by level explosion of finished products and spare parts requirements. At each level the requirements should be netted against available inventory, work-in-process, and planned orders. The actual requirements should be batched according to the rules specified by the users. The netting and batching of components should be suspended until the lowest level at which a particular component may be used, is reached. The part master file contains a record of the lowest level for the item, and determines the stage at which netting and batching should be performed. Thus the gross requirements for components used in a number of product structures can be accumulated before the orders are planned.

2 It should also be possible to explode the production schedule without taking account of available inventory and work-in-process. This gross breakdown is useful for preparing long term plans, and for determining the total requirements for individual part numbers. Such gross breakdown can be carried out at monthly intervals.

3 The system should carry out a clean sweep breakdown i.e. it should generate a new set of orders and requirements ignoring any previous orders which have been planned but not released to the shop-floor. However such breakdowns should take account of the inventory and work-in-process for netting purposes. Clean sweep breakdown is used when a new production schedule is input to the system. The old planned orders are initialized and the requirements generated for both old and new production plans. This procedure should be used if there are a number of requirements which have to be planned, and where it would be better to carry out a clean sweep breakdown in preference to the requirements alterations type of processing.

4 It should be possible to use the system in the requirements alterations mode thereby eliminating the need to regenerate total requirements. Thus any changes to the production plan can be input to the system and their impact on lower level items is rapidly determined. This system mode can be used to update the production schedule and generate the necessary changes to material requirements at random intervals.

5 If necessary the identity of all or selected customer/production orders should be retained during the parts explosion and requirements planning process. This 'pegging' facility allows the management to exercise better control over the production plan and work-in-process.

6 Another desirable feature is the availability and use of inter-active requirements planning techniques which make it possible to review and adjust planned orders at each level.

7 When it is necessary to order items in particular batch sizes the system should include the facility to retain values of confirmed order quantities.

8 Scrap allowances or shrinkage factors should be taken into account in the calculation of net requirements at each level. Similarly safety lead time should be used, in addition to the normal lead time, to calculate the dates on which individual orders should be released.

9 The system should consider the dates on which any proposed engineering changes become effective before arriving at the total requirements for individual components and associated due dates.

10 It should be possible to use different lot sizing techniques according to the rules specified by the management.

11 The input to the system can take the form of a production schedule prepared by the management, sales forecasts or special customer orders. It should, therefore, be possible to explode special customer jobs to determine the required components.

12 Finished products in the production program affected by material or component shortages should be listed out in an exception report so that the management can take the necessary expediting action.

13 The system should include the capability to change from shop days to calendar days and vice versa thereby providing more meaningful reports.

14 It should be possible to assess the financial implications of the total

production schedule as well as individual items included in the production program.

15   Finally it should be possible to use the requirements planning system in the simulation mode so that the impact of possible and desired changes can be assessed in advance of their implementation.

Information Flow in a Requirements Planning System

Fig. 9.2 shows a simplified schematic representation of information flow in a requirements planning system.

The input to the system is in the form of (a) a master production schedule prepared by the management, (b) changes to the production schedule, or (c) any special customer orders. It can be seen from Fig. 9.2 that there are two separate modules used by the requirements planning system. The net requirements for finished products, period by period, are calculated by taking into account the

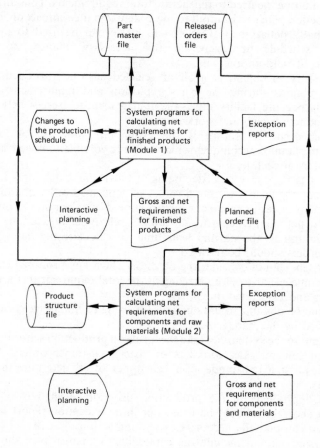

Fig. 9.2 Information flow in an interactive requirements planning system

currently held inventory and work-in-process. This module uses the information, stored on the part master file and released order file, relating to the quantity of currently held inventory as well as any planned and released orders. The output from this module is used as the input to the second module in which the gross and net requirements, for component and raw materials, level by level and period by period, are then determined. This second module determines the gross requirements for components and raw materials by performing a 'Bill of materials explosion'. At each level the net requirements are calculated by subtracting the on-hand and on-order quantities from the gross requirements. Therefore this module needs access to the part master and product structure files. The gross and net requirements can be stored in the part master file itself or its extension. The planned orders can be based on the actual requirements for components or it might be decided to calculate the order quantity using a number of lot sizing techniques. The 'lot sizing' method to be used for a particular item can be specified by means of a parameter in the part master record.

The output from both these modules should show, period by period, the gross and net requirements for finished products along with recommendations for necessary production and purchase orders. In addition the output should include exception reports which highlight conditions requiring personal attention of and action on the part of management. In general the requirements planning system should be able to generate reports in the following modes.

1   Gross and net requirements for all items.
2   Enquiries for individual items.
3   Items for which production or purchase orders have to be placed.

Calculation of Net Finished Products Requirements

In the first module of an overall requirements planning system gross to net calculations for finished products are carried out. The input to this module comprises part numbers of the end items, quantities required, and the due dates. Alternatively the input might show the planning periods, in place of the due dates, in which the finished items are required. The gross requirements in successive time periods for any item, are reduced by the available free stock and any released order quantity not needed to cover previous requirements. The period by period net requirements, calculated above, can be stored on the part master file. At this stage the appropriate lot sizing technique can be used, by calling in the program or routine required to calculate order quantity according to rules specified by the management, so that requirements in successive time periods can be batched together. In addition the main program should include an option to increase the planned order quantity by the appropriate scrap percentage/shrinkage factor stored in the part master record for the item. The manufacturing start date can be calculated by subtracting the item lead time from the due date. If necessary the safety lead time should also be used to arrive at the manufacturing start date. If it is determined, during the offsetting process, that requirements exist at a date prior to the current date then an exception report should be produced. The manufacturing due date for these items can also be determined at this stage to allow time for packing, shipping, and preparation of the required paper work. The system should also include a facility to compare the released production/ purchase orders with the gross requirements. If the released order quantities are inadequate to satisfy the requirements then an exception report should be printed to help the management take appropriate action. The management might expedite orders, cancel them or alter due dates.

In the case of companies which have a relatively simple product range or where both modules of the requirements planning system have not been developed and implemented, the output from this module will be in the form of a report showing, for all items which require attention, the gross requirements, net requirements, planned order quantities arrived at using lot sizing techniques, and their due dates. It is highly desirable that both modules of the requirements planning system should be used to achieve the real benefits of an electronic data processing system. This is particularly true in the case of companies which have a complex product range involving a number of breakdown levels. The output from the first module, showing the part numbers of the finished products, net requirements in each of the time periods or actual due dates in some cases, can be input automatically to the second module.

Calculation of Net Component Requirements

The net requirements and associated due dates for end items, calculated using the first module, are used as input to the second module which calculates the net component requirements by retrieving bill of materials information from the product structure file, and the inventory information from the part master file. The required data processing involves the following operations.

The record of the item for which any requirements exist is retrieved from the part master file. The disc address at which the bill of materials information, relating to any required item, is stored is determined from the part master record. The low level code of the required end item is also stored in a temporary file. This information can then be used to calculate the gross requirements for components assembled to produce the end item. The starting dates for the required quantities can be calculated by subtracting the lead time from the due date. These gross requirements are allocated to appropriate time periods. If necessary a safety lead time factor can also be used during the offsetting process to ensure the availability of the item. The netting and batching should not be performed until the lowest level at which the item is used is reached. At each explosion stage the level code in the product structure file should be compared to the low level code for the item. Once the lowest usage level for the item has been reached the required gross to net calculations can be performed by employing the technique used for finished items. These net requirements, by time periods, can then be stored in the appropriate planned order records and subsequently used to generate the gross requirements for components at the next level. This process is continued until the net requirements for components at the lowest level have been reached.

The net components requirements report should include all the data required for management information and action. Consideration should be given to the inclusion of following data items; gross requirements, net requirements, on hand inventory, in-process inventory, planned order quantities for each of the time periods in the planning horizon. In addition exception reports requiring immediate management action, as in the case of components for which the parts explosion and offsetting process reveals a requirement prior to the current date, should also be produced.

Program Routines

The following program routines are common to the two modules used for the calculation of net finished product requirements, and net component requirements:

1 Calculation of planned order quantities, also referred to as lot sizing.
2 Offsetting to determine starting dates.
3 Processing of amendments to requirements planning system.
4 Interactive planning by interrupting processing at specified levels.

The essential features of these program routines are now considered in detail.

Lot Sizing

The calculated net requirements, over future time periods, can be combined into the order quantity according to the lot sizing policy to be used for the item. The order quantity required to cover any existing and anticipated requirements will vary according to the type of item involved. For example very commonly used items can be manufactured and ordered in quantity so that they are never out of stock. Standard components used on a number of different types of machines might be manufactured or ordered in economic batch quantities which minimize the total manufacturing and carrying costs. However these techniques should not be applied to expensive manufactured items. It might also be desirable to include an option in the system so that the requirements for particular part numbers are never batched together. This will obviously result in a separate order for each individual requirement.

The following policies can be used for individual items:

1 The order quantity can be the actual net requirements in a particular period, and no batching rules are used.
2 The order quantity can be the net rquirements accumulated for a time period specified by the user. The time period over which the demand is batched might be the lead time for the item.
3 The order quantity can be a fixed order quantity specified by the user and stored in the part master file.
4 The actual order quantity might be a multiple of the fixed order quantity. Thus if the fixed order quantity is 30 units and the actual requirements call for 45 units, then the order placed might be for 60 units. The maximum/minimum multiple order quantities are usually specified by the user.
5 The planned order quantity can be calculated using the standard economic order quantity relationship. The classical economic order quantity (EOQ) model attempts to minimize the total cost of producing or purchasing an item, and is based on the following assumptions.

(a) Annual demand rate is fixed and constant.
(b) The item lead time is exactly known.
(c) Shortages are not permitted.

The economic order quantity, Q, is calculated using the relationship

$$Q = \sqrt{\frac{2AD}{IC}}$$

where A = Fixed ordering costs.
$\quad\quad\quad D$ = Annual demand rate.
$\quad\quad\quad I$ = Annual inventory carrying cost rate.
$\quad\quad\quad C$ = Unit production or purchase cost.

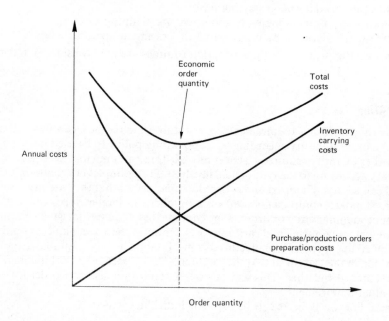

Fig. 9.3 Illustration of the classical economic order quantity relationship

Fig. 9.3 shows the variations in the inventory carrying cost and the production/purchasing costs for different order sizes. The economic order quantity is the point at which the inventory carrying cost is equal to the production/purchase order set up cost, and the total annual cost is a minimum at this point. The total cost curve is relatively flat around the economic order quantity point and may be rounded up or down to a more suitable order quantity.

6   The relationship in the previous section has been used since the early twenties. However the assumptions made in arriving at the economic order quantity very rarely hold in real situations. Better results can be obtained using an alternative approach in which the actual demand pattern, and not the average fixed annual demand rate, is taken into account to minimize the unit ordering and carrying cost.

The unit cost can be calculated from the dynamic programming formula;

$$\text{Unit cost} = \frac{A + H(d)}{d}$$

where A = Fixed ordering cost.
$H(d)$ = Total inventory carrying cost.
d = Order Quantity.

The expected cumulative demand in successive time periods is considered and the order quantity which results in minimum unit cost is selected. The method can best be illustrated by means of a simple example. Table 9.1 gives the expected product demand for successive time periods.

Table 9.1 Demand pattern during successive time periods

Period	1	2	3	4	5	6	7	8	9	10	11	12	13
Demand	40	50	60	70	45								

It is assumed that;

Fixed ordering cost = 5
Unit purchase cost = 1
Inventory carrying rate = 2% per period (0.02)

The first possibility is to order 40 units required to just cover the demand for the first period. The average inventory during the first period is 20 units, and the unit cost is;

$$\frac{5 + (40/2)(0.02)}{40} = \frac{5.40}{40} = 0.135$$

The second possibility is to order 90 units required to cover the demand for the first two periods. The average inventory is $((40/2) + 50)$ units during the first period and $(50/2)$ units during the second period. In this case the unit cost is

$$\frac{5 + ((40/2) + 50)(0.02) + (50/2)(0.02)}{90} = 6.90/90 = 0.0767$$

The next possibility is to order 150 units required to cover the demand for the first three periods. The unit cost is

$$\frac{5 + ((40/2) + 50 + 60)(0.02) + ((50/2) + 60)(0.02) + (60/2) \times 0.02}{150} = \frac{9.90}{150} = 0.0660$$

If the quantity ordered covers the requirements for the first four periods then the unit cost is

$$\frac{5+((40/2)+50+60+50)\ (0.02)+((50/2)+60+50)\ (0.02)+((60/2)+50)\ 0.02+(50/2)\times 0.02}{200}$$

$$=\frac{13\cdot 40}{200}=0.0670$$

The unit cost in this particular case is higher than that for the previously ordered quantity. Hence the optimal order quantity, with a minimum unit cost of 0.066, is 150 units.

By repeatedly using this dynamic programming algorithm the optimal order quantity suitable for any demand pattern can easily be calculated. Although this procedure can be continued to cover the demand for all future periods, in practice it might be decided to examine a situation in which the order quantity minimizes the cost requirements for the first few periods.

7   The company might decide to use a different lot sizing technique which can take account of any quantity discounts or price breaks which might be available. In such a method an attempt is made to determine the order quantity which minimizes the unit cost for meeting the requirements over the planning horizon. The method can best be illustrated by means of a simple example. Table 9.2 shows the price breaks used to meet the expected product demand pattern indicated in Table 9.1.

Table 9.2   Price breaks available for meeting demand pattern in Table 9.1

Quantity	$0 \le d < 50$	$50 \le d < 100$	$100 \le d$
Unit cost	1.00	0.90	0.80

From Table 9.2 it is seen that the direct purchase cost of each item is 1.00 for order quantities of less than 50 units. If the order quantity lies between 50 and 99, the cost of each unit drops to 0.90. The final price break occurs when the order quantity is greater than, or equal to 100 units. In this case the unit cost is 0.80.

The unit cost for an order of 40 units, required to cover the demand for the first period, is

$$C_1 = 1.00 + \frac{5.00+(40/2)\ (1.00)\ (0.02)}{40} = 1.135$$

The unit cost for an order of 90 units required to cover the demand for the first two periods is

$$C_2 = 0.90 + \frac{5.00+((40/2)+50)\ (0.90)\ (0.02)+(50/2)\ (0.90)\ (0.02)}{90}$$

$$= 0.90 + \frac{5.00+1.26+0.45}{90} = 0.90+0.074\ 55 = 0.974\ 55$$

The unit cost for an order of 150 units, required to cover the demand for the first three period is

$$C_3 = 0.80 + \frac{5.00 + (40/2) + 50 + 60)\ (0.80)\ (0.02) + ((50/2) + 60)\ (0.80)\ (0.02)}{150}$$

$$+ \frac{(60/2)\ (0.80)\ (0.02)}{150}$$

$$= 0.80 + \frac{5.00 + 2.08 + 1.36 + 0.48}{150} = 0.859\ 46$$

The unit cost for an order of 200 units, required for the first four periods is

$$C_4 = 0.80 + \frac{5.00 + ((40/2) + 50 + 60 + 50)\ (0.80)\ (0.02) + ((50/2) + 60 + 50)\ (0.80)\ (0.02)}{200}$$

$$+ \frac{((60/2) + 50)\ (0.80)\ (0.02) + (50/2)\ (0.80)\ (0.02)}{200}$$

$$= 0.80 + \left(\frac{5.00 + 2.88 + 2.16 + 1.28 + 0.40}{200} \right) = 0.8586$$

The unit cost for an order quantity of 260 units, required for the first five periods, is

$$C_5 = 0.80 + \frac{\begin{bmatrix} 5.00 + ((40/2) + 50 + 60 + 50 + 60)\ (0.80)\ (0.02) + ((50/2) + 60 + 50 + 60)\ (0.80)\ (0.02) + \\ ((60/2) + 50 + 60)\ (0.80)\ (0.02) + ((50/2) + 60)\ (0.80)\ (0.02) + (60/2)\ (0.80)\ (0.02) \end{bmatrix}}{260}$$

$$= 0.80 + \frac{5.00 + 3.84 + 3.12 + 2.24 + 1.36 + 0.48}{260} = 0.80 + 0.061\ 69 = 0.0861\ 69$$

The unit cost in this case is higher than the unit cost for an order quantity of 200 units. It would therefore be advisable to place an order for 200 items since the unit cost is minimized. This process can be continued till the end of planning horizon and an order placed for a quantity which minimizes the overall unit cost.

It is only the availability of a computer which makes it possible to carry out these relatively simple but repetitive calculations, and achieve the benefits of price break opportunities, to satisfy a particular demand pattern for a large number of items. It also enables the maagement to use different types of lot sizing and ordering techniques for various items.

Within the part master file the batching policy to be used for an individual item is specified. The first period in which the actual requirements for the item exist is used to generate a planned order. The lot size can be calculated using one of the techniques discussed above, by calling in the appropriate program routine. If net requirements are greater than the order quantity, it will be necessary to generate multiple planned orders. All net requirements are added and planned orders generated until net requirements fall to zero. It might be desirable to include a scrap allowance or shrinkage factor and increase the planned order requirements so that the quantity required complete can be produced. The following relationship can be used to calculate the starting quantity.

$$Q_1 = \frac{Q}{1 - SA}$$

where Q = Required quantity
$\quad\quad Q_1$ = Start quantity
$\quad\quad SA$ = Scrap allowance (shrinkage) factor

The lot size determined for one level is then used as input to the next level for parts explosion and netting application. The output from this routine consists of a series of planned orders, order quantities and their due dates for all the required items. In addition exception reports should be generated for items which require management attention.

Calculation of Start Dates (Offsetting)

Offsetting refers to the techniques used to calculate the date on which procedures for procuring the net requirements must be started off. The relevant production or purchase lead time for the item, stored in the part master file, is referenced and used during the offsetting process.

The due date for the planned order requirements for the finished product is offset by taking into account the relevant lead time. The planned order quantity at any level is also the gross requirement for its constituent parts, and the starting date at any level is also the calculated due date for the lower level components. The start date for components at this lower level can be calculated by lot sizing them and subsequently subtracting their lead time from the calculated due date. This process is continued until the requirements for the items at the lowest level, for example raw materials and purchased parts, have been netted, and planned orders raised. The output from this program will consist of a report which shows the quantity and due date requirements for all the items required to satisfy the production schedule. In addition exception reports should be generated for items which have requirements prior to the current period or date. This information is also stored in the planned order records.

In the following simple example, used to illustrate the offsetting process, it is assumed that the finished product F is assembled from a sub-assembly S and two purchased parts P_1 and P_2. Sub-assembly S is manufactured in house using components A and B. Materials C and D are required in the production of items A and B respectively. Thus the product structure is as shown in Fig. 9.4.

It is assumed that the requirements for item F in weeks 34 and 36 amount to 30 and 40 units respectively. There is no requirement for item F in week 35. It is further assumed that the quantity on hand for product F is zero, and it is management policy to manufacture this item against specific customer orders only. The start date (or period) can be calculated by offsetting the assembly lead

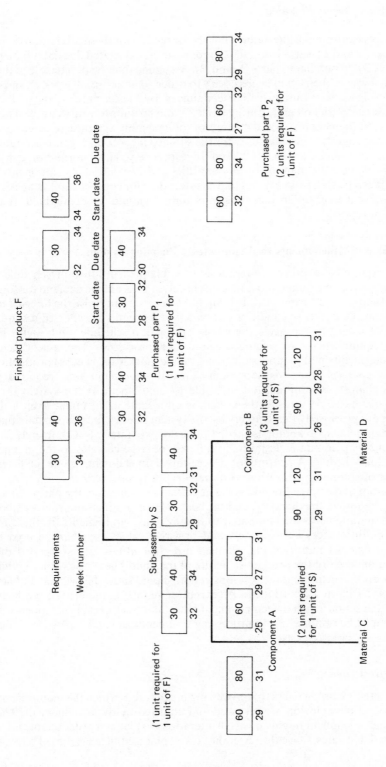

Fig. 9.4 A simplified representation of a requirements planning system

time of 2 weeks from the due dates. Thus the start dates for these two requirements are weeks 32 and 34 respectively which represent the calculated due dates for sub-assembly S and purchased parts P_1 and P_2. Assuming that the purchase lead time for P_1 is 4 weeks, then it will be necessary to order 30 units during week 28 and 40 units during week 30. Similarly for a lead time of 5 weeks for purchased part P_2, it is necessary to order 60 units during week 27, and 80 units during week 29. There is also a requirement for 30 and 40 units of sub-assembly S at the end of weeks 32 and 34 respectively. The start dates for the sub-assembly S, taking into account a lead time of 3 weeks, are weeks 29 and 31 respectively. This information can be used to calculate the planned order quantities and due dates for components A and B. It can be seen from Fig. 9.4 that the start date for component A is week 25. If the current week is 26 then this information should be highlighted as an exception report.

Processing of Amendments to Requirements Planning System

The capability to process alterations to a planned production schedule is necessary since a customer may cancel his order, or increase the demand for some items at short notice. Alternatively it might be necessary to revise the forecast for some items. Under these conditions when only a few items are affected it should not be necessary to reprocess the whole production schedule. Only items for which requirements have changed should be processed. The amendments processing facility makes it possible to determine quickly the effect of production schedule changes input to the system. Since the computer time required to process changes is minimal, it is possible to input them to the system on a continuous basis. As a result the system is up to date at all times. Even if the changes are processed at daily intervals, the system is much more up to date than if new requirements planning is carried out at weekly intervals. Completely new requirements planning is desirable in the case of large changes to the planned production schedule. A continuous processing of amendments, while inefficient in terms of computer utilization, is more effective in user terms.

The input to the amendments processing program consists of the part number, quantity required, the due date, and the customer or production order number. The new requirements are compared to the previous requirements for the same item. The differences between new and previous values can then be used to calculate new net requirements, planned order quantities and associated due dates for all levels of the product structure of the item. These new values update the appropriate fields, and the resulting report should show the original and new values of the affected record fields. An exception report should also be produced to show the action which the management must take. Such exception reports will, for example, be required if the processing of amendments affects orders which have already been released.

Interactive Planning

Interactive planning refers to a processing procedure in which the requirements at each level of explosion are reviewed and adjusted by the management. This review and adjustment might consist of alterations to planned order quantities or associated due dates. Once the review has taken place, adjusted values of relevant parameters, i.e. order quantity and/or due date, can be input to the system and the parts explosion to the next level continued. If necessary this procedure of interruption and review may be repeated at a number of levels after the processing for that particular level has been completed. The level or levels at which processing should be interrupted can best be specified by means of a suitable user input parameter value. Interactive planning is a useful tool which enables the management to interact with the system and use its intuition and other knowledge of the existing situation. Best results are obtained by using this interactive planning facility in an on-line mode since all the changes can be rapidly processed, and the frustrating delays and errors inherent in the batch processing mode are not encountered. In on-line as well as the batch mode of operation it will be desirable to produce an exception report which shows the differences between orders as planned by the system, and the changes made to them after the management review.

Preparing for the Requirements Planning System

A requirements planning system can make a dramatic contribution to the success of a company provided that proper use is made of it and the necessary disciplines exist. Only a few companies have managed to establish effective requirements planning systems in spite of the efforts of a large number of companies. Very few requirements planning systems have attained the desired objectives, but the successful systems have resulted in very considerable benefits. One of the main causes of failure is the lack of appreciation by the users of the fact that the requirements planning system is much more than the parts explosion process combined with the use of re-order point techniques. A requirements planning system is a new approach to running a business which lays a very great deal of emphasis on formal planning and formal procedures combined with accurate recording of inventory transactions as and when they occur.

The implementation of an effective requirements planning system is a major challenge to the management of any company and requires commitment of large amounts of time and money. It must be properly planned for after the decision has been made to go ahead with the project. The overall charge of the implementation of a requirements planning system should be in the hands of a person who believes in requirements planning and can see the long term benefits because the project duration is longer than most people anticipate even if all the required software is available. Many production/manufacturing managers are trained to deal with day-to-day problems and are impulsive by nature. Left to such people, the requirements planning system will never become really effective because in addition to converting the system, it is necessary to make sure that the attitudes of people using the system also change. It is a serious mistake to implement a requirements planning system when the managers and other users still think in terms of independent demands. The users themselves must be actively involved in the project. The staff should be informed that a requirements planning approach is going to be used for running the business so that they do not maintain informal systems in parallel with the formal system. It is only then that they will appreciate the facilities available and how to use them.

For the implementation of an effective requirements planning system, it is necessary to concentrate on all the modules which interface with this system.

These are:
1   Bill of Materials.
2   Master production schedule.
3   Inventory recording.

If the bill of materials is not well structured and up to date taking into account engineering design changes, then the requirements planning will not result in any significant benefits. The bill of materials structure must conform to the actual assembly process, otherwise the system will reveal requirements at wrong times.

The master production schedule is the prime input to the requirements planning system and as such forms the key to its success. Above all this schedule must be realistic and should be subjected to frequent management review. An unrealistic and inaccurate master production schedule will only result in plans which cannot be implemented on the shop floor. Even minor errors in the master production schedule can cause very serious requirements planning problems.

Accurate inventory records are also critical for the success of a requirements planning system. A requirements planning system used in conjunction with inaccurate inventory records will never be effective because even a realistic production schedule, based on accurate demand forecasts, will reveal requirements which do not actually exist. Conversely it might show that requirements can be satisfied from available free stock when in fact stock has fallen to a very low value. This of course reveals the need for perpetual inventory counts discussed in a following chapter.

An effective requirements planning system must have access to accurate lead time data. These requirements for accurate detailed data and the strict discipline and control procedures result in an increase in the duration of a requirements planning system project. However, they are absolutely necessary if the requirements planning system is to be a working reality.

The management as well as the users should be educated about the concepts used in the requirements planning system and the anticipated impact of the system on the company profits and on the procedures used in the company. This educational aspect is very important and should not be overlooked.

Summary

With a computerized material requirements planning system, more detailed planning than would be possible with manual systems can be carried out. Exception reports can be produced to bring salient facts to the attention of the management. Small alterations to the production schedule can quickly be taken into account without the need to process the whole of the production schedule. Thus the impact of orders additional to the production schedule can easily be assessed. It is also possible to determine end items affected by the shortage of materials or components at lower levels. This enables the management to either inform the customer about delayed deliveries or assign a higher priority to the manufacture of affected components so that they can be delivered on time. Interactive planning can be used to interrupt processing at a number of stages and thereby allow the management to review and rapidly adjust the requirements at any stage during the parts explosion process.

The high speed of the computer can be used to include in the system a variety of sophisticated lot sizing techniques, requiring relatively simple but repetitive calculations.

10
Capacity Planning System

Introduction

For the implementation of an effective production planning and control system it is necessary to consider the availability of not only components and materials but also the required manufacturing capacity. In this context manufacturing capacity refers to the availability of machines as well as the manpower. In most manufacturing organizations there is a major problem of trying to balance the available machine and manpower resources required to manufacture all the items included in the production schedule. Such balancing is required in order to reduce the manufacturing lead time, the capital invested in the work-in-process, and queues at certain work/machine centres, which result in late orders. While it is impossible to match, at any particular instant of time, the available and required capacity, it is nevertheless possible and highly desirable to carry out detailed capacity planning. If it is discovered that the available capacity is inadequate to meet the planned requirements, then the management can take the remedial action required to increase the capacity, by sub-contracting, overtime working, or working two shifts in critical areas. Conversely in the case of a decrease in orders received any desirable reductions in manpower capacity may be catered for by natural wastage. It might also be possible to reduce bottlenecks in some areas by shifting personnel from one area to another.

If enough capacity is available for performing all the required production operations, then all the jobs can be processed immediately after they are received at the appropriate work centre. Such immediate processing is rarely encountered since there are always machine and/or manpower capacity restrictions and other jobs also require the use of the same capacity. This results in a queue of work waiting to be processed. Overtime working at a particular work centre is not always helpful since a large amount of work processed here only results in the development of a similar queue at the next work centre and the requirement to work overtime to relieve this queue. It is also necessary to increase the safety stock level due to the increased lead time.

The early release of a planned order, by increasing the production lead time, does not solve any problems. In fact it only increases the volume of the work-in-process. The queue times at various work centres also increase and orders are still completed behind schedule. The early release of an order, without adequate capacity planning, does not increase the available capacity and there is no improvement in the delinquency situation.

Often the capacity is not adequately planned for, and orders are released onto the shop-floor in the hope that any work input to the system will eventually appear at the other end. In practice this results in large amounts of back-scheduled work, long queues and delays at some of the work/machine centres, excessive expediting, and increasing amounts of overtime leading to ever increasing costs. This is not to imply that capacity planning can be easily carried out since the capacity is rarely fixed. The capacity is reduced as machines break down, key workers are absent, or as the efficiency of production varies. Conversely the capacity can be increased by working overtime or extra shifts. The capacity also varies from one period to another according to the length of the shift, manpower assigned to the work centre, amount of overtime being worked, and statutory or customary holidays which might fall during the period under consideration. Another factor which introduces considerable complexity into the capacity planning function is the fact that there are a large number of independent production orders which are progressing, at any time, through available production facilities. Thus before releasing further orders onto the shop-floor, the management has to ensure that facilities required to manufacture these items are available. The problem of capacity planning is particularly difficult in the case of companies with a wide product range. The fluctuations in the product mix, the volume of various products to be manufactured, and the complexity of routes followed by these products through a maze of production facilities, increase the difficulties encountered by the production planner in assessing the impact of production schedules. One simple and obvious solution is to ensure that available capacity is always in excess of the required capacity. However in practical terms such a solution is rarely possible. The other alternative is to carry out detailed capacity planning. The availability of a digital computer makes it possible to perform this planning in an interactive manner until the required and available capacities are balanced.

A capacity planning system is concerned with long term planning of the resources. These manpower and machine resources cannot usually be altered at short notice. Therefore the management has to determine or forecast its future requirements over the planning horizon time period, so that it is possible to make judgments and take necessary action early enough to be able to adjust the future capacity in line with the projected requirements. The planning horizon varies from one manufacturing company to another depending upon its product range. A company engaged in the production of aero-engines, for example, will have a much longer planning horizon than a company manufacturing simple consumer goods. Typical planning horizon time periods might be one, three, six, twelve or more months.

The capacity planning is usually carried out by calculating the starting dates and loading the projected requirements, at scheduled times, against the available machine and manpower capacity which makes it possible to establish the required load pattern. Capacity planning systems are often confused with the related operations scheduling and sequencing system. An operations scheduling system is concerned with short term problems such as the batches of work on which operations should be performed on a given day, and the relative priorities of various work batches. In contrast a capacity planning system is not concerned with individual jobs but relates to the total load on a given work centre over a total period of time. As a result the capacity planning system is a rough estimate of total activities whereas operations scheduling is a more exacting process. The major advantage of a long range capacity planning system is that it helps the management develop a levelled load on various work centres. In addition such a system can be used to reduce the product lead time by minimizing the waiting time at various work centres, and help the management ensure that men and machines are not idle. The available operators can be shifted from one work centre to another in order to meet the requirements.

A computer-based capacity planning system can also be used as an excellent simulation tool. The possibility of meeting the requirements by addition of extra capacity, overtime working, or by allocating higher priorities to certain orders can be rapidly examined. Similarly the impact of some large orders on the required capacity can be assessed before actually releasing the order, and the management can take any required corrective action. The implications of various possibilities in terms of manpower requirements, product costs and work schedules can be quickly calculated. Also the possibility of balancing the required and available capacity, by altering the due dates for certain orders, can be determined. It would be impossible to carry out such exercises in the absence of a computerized capacity planning and simulation system.

Principles of Capacity Planning

To optimize the use of available resources and satisfy the production schedule requirements it is necessary to consider the total manufacturing load, the sequence of operations, i.e. the order in which operations on particular batches must be carried out, and the available machine and manpower capacity, usually measured in terms of machine hours and man hours respectively.

For the batch manufacturing operations under consideration the best approach is to simulate the capacity needed to meet the production schedule requirements using the planned order, routing, work centre, and work-in-process files. The planned order records contain information with regard to the items to be manufactured during the planning horizon time period. The routing file indicates the sequence of operations which must be performed, in a number of work centres, to produce the item. For each operation the set up time, run time, average queue time, and the time taken to move a standard amount of work to the next work centre are also listed. The information relating to the potential availability of individual machine/work centres is stored in the work centre file. In some cases in which the work usually performed in a particular work centre can also be carried out, albeit inefficiently, in other work centres, in case of overload conditions etc., this information is also stored in the work centre file. Where routine maintenance is carried out as part of the working shift it is necessary to indicate the level of availability of work centres on particular days. A shop calendar can be used to show this availability.

A number of different principles can be used for capacity planning purposes. Using the 'infinite capacity forward scheduling principle' the work on a given batch is started on the specified date or the current date if no starting date is specified, and the sequence of operations is loaded onto the relevant work centres taking into account the processing, queuing and transit time for each operation. The alternative system of infinite capacity backward scheduling is based on the assumption that all the items will be produced on the due date. The time taken to carry out the last operation is loaded onto the appropriate work centre and the start date for the final operation calculated. Once again the queuing and transit time periods are fully taken into account. The same procedure is repeated for all the operations on the routing file which must be performed to manufacture the item.

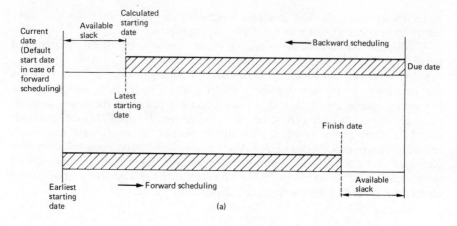

Fig. 10.1a Forward and backward scheduling principles

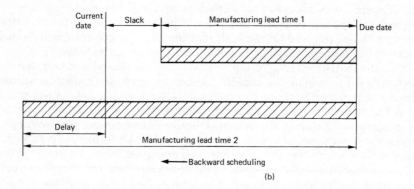

Fig. 10.1b Delay and slack in backward scheduling

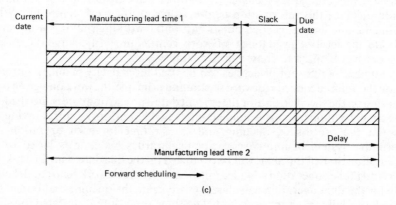

Fig. 10.1c Delay and slack in forward scheduling

If the final starting date is before the current date, then it is necessary to reschedule the work in the forward direction from the current date and a new due date must be calculated. The delay may be overcome using expediting procedures. Fig. 10.1a illustrates the forward and backward scheduling principles.

The slack or delay associated with an order can be calculated after it has been scheduled. In forward scheduling the slack is the time available between the calculated finish date and the order due date. In backward scheduling the slack is the time available between the current date and the start date.

Similarly, in forward scheduling the delay refers to the time between the due date and the date on which the work is completed. In backward scheduling the delay is the time interval between the start date and the current date, which must be made up by expediting etc. if the orders are to be delivered on time. Figs 10.1b and 10.1c illustrate the slacks and delays associated with backward and forward scheduling.

The capacity load profile, showing the underloading and overloading of various work centres, can then be calculated. Programs to smooth out the load are used to match the load against available capacity. These programs are based on the principle of finite capacity scheduling, i.e. operations are loaded as and when capacity becomes available. The output from these programs also indicates the delays on various operations. Priority codes can then be allocated to urgent orders which must be completed on time. The programs can then be rerun to assess the effect of altered priorities. In general it is desirable to ensure that priority is not allocated to more than 5% of the total load. In some cases the management might wish to examine the effect of splitting batches, using alternative work centres or production paths. In all cases alternative program routines can be used to examine the implications of various decisions.

Fig. 10.2 shows the flow of information in a typical capacity planning system.

Capacity Planning Calculations

The input to the capacity planning system consists of planned and released orders, and the required operations are then loaded by reference to the routing file which also shows the work centres at which the individual operations are carried out.

The first step in a capacity planning system is the establishment of the time period over which individual operations are to be carried out. This involves the calculation of the required due dates and start dates of the individual operations, for all the orders, scheduled over the planning time horizon. For the preparation of a realistic production schedule it is important to ensure that in addition to the time required to carry out the necessary operations, adequate allowance is also made for the transit time between operations, and the time spent waiting for the appropriate machines and necessary tools to become available, i.e. the queue time. In fact the actual processing time is a small proportion of the total time between the due date and start date. Although the inter-operation time is usually fixed, taking into account all the relevant factors involved, it is only a rough indication of the time which should be allocated for completion of the operation. In practice the actual time taken to complete the operation may be substantially

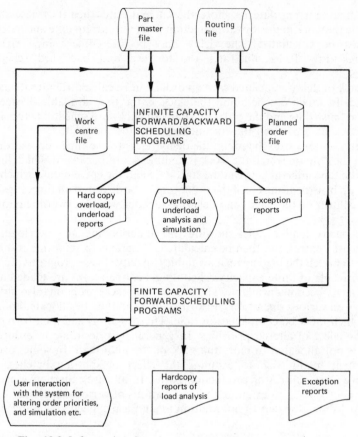

Fig. 10.2 Information flow in a typical capacity planning system

different from the allowed time depending upon the total number of jobs which have to be processed, the product mix, the queue at the work centre, and priority allocated to individual orders.

The overall time required to carry out an operation consists of the following five separate elements.

1   Queue time prior to processing, Q_p

2   Estimated set up time, S_t

3   Standard run processing time, R_t

4   Queue time after processing, Q_a

5   Transportation time, T_t

Fig. 10.3 graphically illustrates the overall operation time.

Queue time before processing is the average time spent waiting in a work centre before the work on a batch can be started.

Estimated set up time is the time allowed to prepare/set up the machine so that the operation may be performed. The set up inspection time is also included in this element. Set up time is independent of the quantity to be processed, and for some operations it may be zero.

Standard run processing time refers to the time allowed for actual performance of the specified operation, for example grinding or fitting. The time required to carry out an operation on a given order quantity is calculated by multiplying the standard run time and order quantity.

Queue time after processing is the average waiting time before the batch of work is picked up for transportation to the next work centre or another area in which batches of work are stored between operations.

Transportation time is the average time needed to transport the order from one work centre to another, and depends upon the distance between them. In some systems the queue time after processing is combined with the transportation time.

In practice it is often necessary to reduce the time required to carry out an operation. This can be achieved by expediting and reducing the queue times as well as the transportation time. For example following a reduction of 20% in the queue and transportation times, the total time required to perform the operation mentioned earlier is as shown in Fig. 10.4.

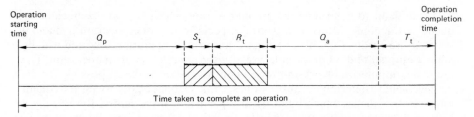

Fig. 10.3 Elements of total time required to complete an operation

The time spent in the queue usually varies depending upon the overload or underload condition of the particular machine or work centre. The queue times usually determine the manufacturing lead time. It is not unusual for the queue times to form up to 80 or 90% of the manufacturing lead time. Therefore the total time taken for an operation can be divided into two elements, as shown in Fig. 10.5.

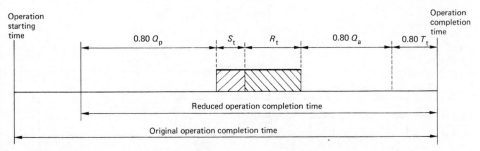

Fig. 10.4 Reduced operation completion time following a 20% reduction in queue and transportation times

F

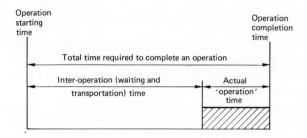

Fig. 10.5 Division of total operation time into inter-operation time and actual operation time

(a) The actual time spent on the machine including the set up time.
(b) Inter-operation time during which the job is moved and is waiting in the queues.

It is difficult to determine the amount of time which should be allocated to queue parameters. Many companies, especially the ones operating manual systems, use a rule of thumb with regard to the inter-operation time and usually allow a number of days or a whole week for the completion of an operation. The use of a computerized work-in-process control system makes it possible for the management to determine realistic values of queue times by observing at frequent intervals, the queues which build up at a particular work centre. By drawing a histogram of the frequency at which a particular queue size (hours of work waiting to be processed) occurs at the work centre, the average queue size can be computed and used in future product lead time, and capacity planning calculations. Also, with such a system, if the queue gets out of control, i.e. there is always a very long queue resulting from a large backlog of work due to machine breakdown etc., this condition will be shown up quickly by an abnormal frequency distribution and the management can take corrective action by a 'once for all' overtime working.

Based on the time required to carry out various operations the total lead time, shown in Fig. 10.6, can be determined and used for calculating the date on which work must start.

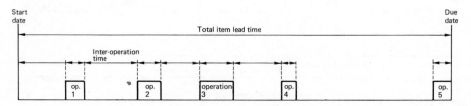

Fig. 10.6 Total product lead time in terms of actual operation times and inter-operation times for all the required operations

Infinite Loading Program

The input to the infinite loading program will typically consist of the following data which also represent the output from the requirements planning system.

(a) Part numbers of items which have to be produced during the planning horizon.
(b) Quantities of items required.
(c) Due Dates for these items.

The infinite capacity backward scheduling program works backwards from the order due date which is also the time at which the last operation is completed. The start date and the actual processing time for the last operation can be calculated by retrieving the operation lead time, set up time and run time from the routing file. The start date of the last operation represents the finish date of the previous operation, for which the start date and actual processing time can now be calculated using the same procedure. This procedure is repeated until the time for the start of the first operation is reached.

During the capacity planning process it is also necessary to consider the operations which still have to be performed on the released orders, i.e. the work-in-process currently on the shop-floor. In a fully developed system it would be desirable to monitor the work-in-process (WIP) using on-line terminals so that the current activity state is fully taken into account during the capacity planning calculations. If an on-line WIP data collection system has not been installed or if the progress of individual operations is not recorded, then it would be necessary to assume that these operations will be performed at the scheduled times. It is therefore necessary that the program should first initialize the resources required over each of the time steps of the planning horizon, by taking account of the released orders. The capacity available for planning can be determined by subtracting the currently released orders load from the normal capacity. In addition capacity allocated to scrapped work or service orders should also be taken into account in the calculation of net capacity, over the planning horizon time periods, for all the work centres. The problem of variable capacity may be overcome by including the normal and maximum capacity data in the work centre records. Changes in normal capacity, due to holidays etc., can be indicated by including an optional input parameter, which shows the percentage by which the capacity should be reduced or increased, in the system program. Then by considering each item, for which future requirements exist as indicated in the planned order files, a list of operations to be performed and their associated time factors can be prepared. The total run time can be calculated by multiplying the order quantity and the unit processing time. The sum of the set up time and total run time gives the total load on the work centre. Where set up times are appreciable and the machines are set by personnel other than the machine operator, it would be desirable to calculate the load in two separate categories, normally the machine load and manpower load. The calculated time can then be added to the load on the resource required to carry out this particular operation. The addition of the required load may or may not result in an overload condition. If it results in an overload then an exception report should be produced. Once all the operations on the first item, for which requirements exist, have been completed the same procedure can be repeated for all the other required items, and infinite load reports for all the affected work centres produced. The load reports can best be produced in a graphical format for easy interpretation by the management. Fig. 10.7 shows a typical load analysis report. In addition a report in the form of a table should also be produced. The overload conditions should be highlighted by means of suitable distinguishing characters such as asterisks.

During backward scheduling a frequently encountered problem is the lack of time required to complete the operations. As a result the calculated start date for some operations may fall before the current date. It is, hence, necessary to include a facility, in the infinite capacity load calculation program, which can be used to reload all the operations starting from the current date, i.e. carry out forward scheduling. Thus a new completion date can be calculated and management informed by means of an exception report. By specifying the percentage reduction in queue times etc. the earliest completion date, possible by expediting action, can be calculated. Alternatively the management might decide to allocate a high external priority to the order or advise the customer about delayed delivery. The logics of the infinite capacity forward scheduling program and backward scheduling programs are very similar. The only difference is that in forward scheduling the first operation is loaded on the specified date or the current date and all the following operations are loaded taking into account the appropriate operation times, until the final operation has been reached.

It is highly desirable that the load analysis programs should have the facility to depict the load in variable time steps. With such a facility the load can be shown for each day, for the first few days, then in weekly periods, and then in four weekly periods. The amount of detailed information decreases for increasing future time periods.

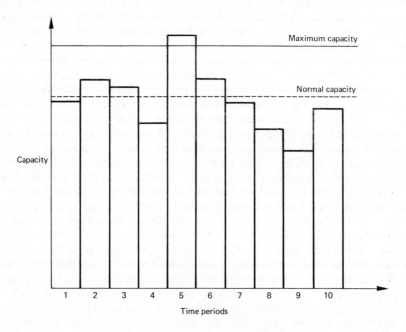

Fig. 10.7 Typical load profile report

Infinite load analysis shows the amount of capacity which should be made available so that the desired customer service levels and inventory policies, as reflected in the requirements planning system, can be implemented.

Shop Load Levelling

Although an infinite load analysis assists the management in the assessment of the capacity required to meet a given production schedule, it does not help in producing a realistic production schedule which takes account of the overloads and underloads at individual work centres during different time periods. In the long term the managment might be able to increase capacity by buying additional machinery or increasing the manpower if either of these decisions can be fully justified. In the medium term management can temporarily increase capacity by overtime working, sub-contracting, or by shifting operators from one work centre to another. However, it is still necessary to prepare a realistic production schedule.

The shop load levelling process can help the management in the development of a realistic production schedule by loading the shop to the actual finite capacity specified in the work centre records. The shop load can be smoothed by shifting jobs backward or forward until the load levels are balanced, taking into consideration the normal or maximum capacity. As a result realistic order start and due dates can be calculated which also helps in releasing orders to the shop-floor in the best possible sequence.

During shop load levelling to finite capacity, the main objectives are to ensure that the manufacturing cycle time is minimized, due dates are met taking into account the actual capacity, and the batch of work does not have to wait for some operations because of capacity bottlenecks. During the infinite loading process the work is scheduled irrespective of whether or not adequate capacity is available. This scheduling is illustrated in Fig. 10.8a. In actual practice adequate capacity may not be available to perform, for example, the third operation because the work centre is overloaded. Thus the completion of the last operation will be delayed by the lead time for the third operation, and the order due date will not be met (Fig. 10.8b). Also, after the third operation has been completed, there may not be enough capacity available to carry out the fourth and fifth operations, in which case the order will be further delayed.

In this case it might be better to release the order early, if possible, so that the due date can be met and the manufacturing cycle time can be minimized (Fig. 10.8c).

It is obviously desirable to shift orders into earlier time periods provided, of course, that such slack time is available and the necessary resources are also available during the earlier time periods. This might help in reducing some of the overload conditions. However, the overload still may not be eliminated. At this stage it might be desirable to shift the load to any alternative work centres. The information relating to alternative work centres, on which similar work can be performed, is included in the work centre record. If overload still persists then it would be necessary to allocate priorities to various orders and carry out a new scheduling run in which orders with higher priorities are loaded before low priority orders. The levelling of over- and underload conditions for work centres,

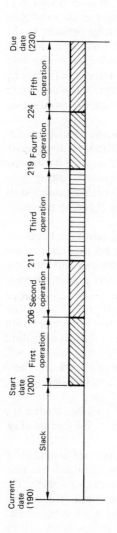

Fig. 10.8a Infinite capacity backward scheduling

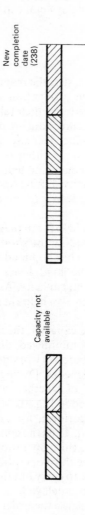

Fig. 10.8b Due date not satisfied due to lack of capacity required to complete operation 3

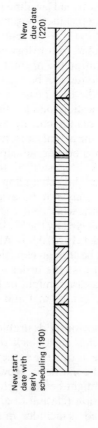

Fig. 10.8c Early scheduling resulting in due date being satisfied and the minimization of cycle time

during various time periods, can smooth out the load variations, but the overall capacity for the whole of the planning time horizon must cover the total load requirements. The finite load analysis report should identify jobs which cannot be completed on time due to the twin constraints of time and capacity. This will either necessitate overtime working or sub-contracting, or result in delayed deliveries.

Order Priorities

The order priorities used during the shop load levelling process, can be calculated in a number of different ways. They can be computed internally within the system or the management may assign individual priorities to urgent orders. It is also possible to use a combination of the two systems whereby the order priority calculated by the system is modified to take account of the user assigned priority values.

The internal, system calculated, order priority can be in the form of a critical ratio defined by the relationship;

$$\text{Internal critical ratio} = \frac{OD - CD}{TR} \tag{I}$$

where OD = order due date
   CD = current date
   TR = time required to complete remaining operations

The critical ratio of less than 1.00 indicates that the order is behind schedule, and must be expedited if it is to be completed by the due date. The lower the value of this ratio, the higher the priority associated with that order. Similarly a critical ratio of greater than 1.00 means that some slack is available on the order.

Let us, for example, assume that the due date for a particular order is shop day 220, current date is 205, and 20 days are required to complete the remaining manufacturing operations. In this case the internal critical ratio is;

$$\frac{220 - 215}{20} = 0.75$$

The order is clearly behind schedule and, therefore, a high priority value is associated with it, when compared to another order with a critical value of, say, 0.90.

The external priorities assigned by the management to different orders can also be based on a suitable scale. The scale may be from 0 to 1, 0 to 10, or 0 to 100. In this case a large value of this parameter indicates a higher priority.

In some cases it is necessary to use the external priority system in conjunction with the system calculated order critical ratio. This facility enables the management to show preference between different orders with approximately equal internal priorities. To avoid the conflict between the systems of management assigned priorities and the internally calculated critical ratios, it is necessary to ensure that an order with a high external priority results in a low value of the critical ratio. Assuming that the external priority is on a scale from 0 to 100, then the overall critical ratio can be calculated using the following formula;

$$\text{Overall critical ratio} = \text{Internal critical ratio} \times \frac{(100 - \text{External priority})}{100} \quad \text{(II)}$$

A default external priority of zero indicates that the overall critical ratio is equal to the internal critical ratio calculated above. If the management wish to attach an external priority of 95 to the above order with the internal critical ratio of 0.75, the overall critical ratio becomes;

$$0.75 \times \frac{(100 - 95)}{100} = 0.75 \times 0.05 = 0.0375$$

Assuming that the management assigns an external priority of 25 to another order with an internally calculated critical ratio of 0.75, then the overall critical ratio becomes;

$$0.75 \times \frac{(100 - 25)}{100} = 0.75 \times 0.75 = 0.5625$$

Since a low value of critical ratio results in a higher priority, the first order will appear higher up in the list of orders to be scheduled.

If the management expedites an order by reducing the queue and transportation times by a specified percentage, the internal critical ratio also changes. In the above example, if, of the 20 remaining days 15 are spent in queueing and transportation etc. and these times can be reduced by a third, then the queue and transportation time is 10 days which when added to 5 days required to complete the remaining operations, gives a total of 15 days. In this case the denominator of equation I decreases and the calculated critical ratio is higher (a value of 1.00) than if the order had not been expedited. It is thus necessary to take account of the expediting factor in the calculation of order priorities. The modified internal critical ratio can be calculated according to the relationship;

$$\text{Modified internal critical ratio} = \frac{OD - CD}{TR} \times \frac{(100 - PR)}{100} \quad \text{(III)}$$

where PR = percentage reduction in queue and transportation times.

Thus the order expediting of 33% results in a modified internal critical ratio of;

$$\frac{220 - 205}{15} \times \frac{100 - 33}{100} = 0.67$$

If the management now assigns this order an external priority of 95, then the overall critical ratio, which applies to all the operations within the order, can be calculated as;

$$\text{New overall critical ratio} = 0.67 \times \frac{100 - 95}{100} = 0.0335$$

The new overall critical ratio is lower than the previously calculated value, without expediting, of 0.0375, and, therefore, a higher priority will be associated with it.

It might be desirable to ensure that the critical ratio never has a value of zero. In such a case the priority calculation program might contain a loop whereby an external priority of 100 results in an overall critical ratio of 0.0001 or some other very small value.

The order priority based on the value of the critical value of the critical ratio is a dynamic quantity which changes every time a new system run is made. An order which had a low priority yesterday, and consequently no operations have been completed on it, will have a higher priority today.

Shop Load Levelling Report

The output from the finite capacity shop load levelling programs can be prepared in the form of a job sequence list of operations, in order of priorities associated with individual orders, which should be performed in a particular work centre. The job of the foreman and progress chaser can be eased to a considerable extent by the inclusion, in the job sequence list, of data such as order due date, operation due date, priority allocated to the order, order quantity, last operation performed on this order and the work centre at which previous operation was carried out. Exception reports can be used to highlight orders unlikely to be processed due to inadequate capacity.

Summary

Simulation of capacity required to meet a given production schedule enables the management to plan in advance of difficulties and problems which would otherwise be encountered on the shop-floor. By taking timely action on underload and overload reports the management can ensure high manpower and machine utilization levels, thereby increasing the production efficiency.

If some work centres are overloaded during particular time periods then some of the orders can be allocated to time periods during which the work centre is underloaded. Alternatively it might be possible to shift work from one work centre to another. All these possibilities can be accurately assessed using the computerized capacity planning system. Similarly the effect of altering external priorities assigned to orders can be quickly determined.

A capacity planning system also helps in reducing the queues at various work centres. As a result the capital invested in the work-in-process is reduced.

By striking a balance between the required and available capacities, a high percentage of orders can be delivered on time without the need for excessive expediting and overtime. Once the system is fully developed and operating effectively then it is also possible to reduce the manufacturing lead time and delivery period. As a consequence the competitive position of the company improves. The potential for increasing the capacity by improving the efficiency levels can also be realized.

11
Order Release and Operations Scheduling

Introduction

The management of manufacturing companies has to exercise adequate control over the release of planned orders to the shop-floor. The order release process forms the interface between the planning phase and the subsequent execution of these plans. Unless the order is released on the scheduled date all the time and money spent in the planning effort is wasted. In addition, once an order has been released any subsequent changes are difficult and expensive. The management must therefore check the availability of materials and components required in the production process before the order is actually released, and any detailed short term operations scheduling calculations are carried out. With most manual systems a commonly used procedure is to physically collect together all the required items before the order is released. This manual staging of materials and components is often time consuming and means that if the order cannot be released due to a shortage of some of the items, the items already collected cannot easily be used for any other order. This particular procedure requires a substantial amount of factory floor space which is frequently at a premium. With a computerized system it is no longer necessary to collect together all the required items. By including suitable fields in the inventory records and pegging materials to particular planned orders, the inventory items can be reserved thereby reducing the effort required on the part of the stores personnel. A well developed order release system can quickly determine the items not available, and a report for management action can be prepared. This report can then be used to expedite the supply of short materials. Once all the items required to start the order become available and the order is ready for release to the shop-floor the necessary works documentation, for example manufacturing route, detailed description of operations to be performed, material picking list, material requisitions, turn-around documents, can automatically be generated by the system. The order release function can best be carried out in an on-line mode.

Following the release of the order it is necessary to plan for the actual performance of required operations on a large number of batches of work currently on the shop-floor. Infinite loading procedure shows the long term capacity required to meet a given production plan so that the management can make the necessary provisions. The shop load levelling carried out in the medium term ensures that the orders are scheduled taking into acccount the actual capacity available.

Some companies, particularly those with comparatively simple product structures, might use the reports produced by the shop load levelling programs for actual loading of different machines. In order to prepare realistic daily production schedules it is necessary to commit machine and manpower capacity explicitly to the orders on which work has to be done. Operations within an order

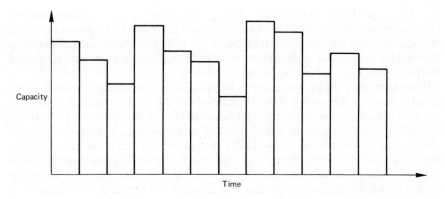

Fig. 11.1a Long term infinite capacity loading (6–12 months ahead)

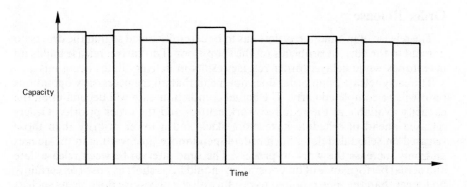

Fig. 11.1b Medium term shop load levelling (1–13 weeks ahead)

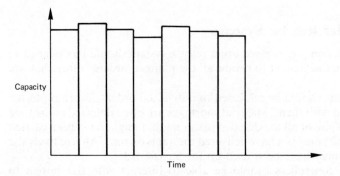

Fig. 11.1c Short term operations scheduling (1–7 days ahead)

are not considered individually during the shop load levelling process. Operations scheduling, which takes into account individual operations, is carried out during detailed machine loading. Fig. 11.1 illustrates the relationship between the time scales of the long term infinite loading, medium term shop load levelling, and short term operations scheduling.

The main objective of operations scheduling is to provide the foreman with a list of orders on which work should be carried out during a predefined period, for example a shift, day, or week. In order to consider realistically any changes in the shop-floor conditions or non-availability of materials etc., it is highly desirable that the detailed operations scheduling system should be run at very frequent intervals, typically at the beginning of each shift. An on-line real-time system is highly benificial for this application so that if the shop-floor operations are no longer under adequate control due to machine breakdowns or absence of key operators the necessary system programs can be rerun to produce a new work schedule.

Order Release

The release of production orders has to be carefully timed so that the orders do not add to the current problems on the shop-floor. Too early a release builds up inventories while delayed order release results in delivery dates being missed.

The early release of an order does not mean that all the necesssary operations on it will be completed early. The actual completion date will depend upon the capacity available at the required work centres and the order priority. Orders released ahead of schedule have more slack and a lower priority than those released on scheduled date. Such orders spend more time waiting in the queues and only increase the work-in-process. The time interval between release date and actual performance of first operation should be as short as possible; certainly no longer than the time required to produce all the necessary documents such as engineering drawings and material requisition slips, pick up all the items and transport them to the work centre at which the first scheduled operation has to be carried out.

Features of Order Release Systems

A comprehensive computer-based order release system should be designed so that it enables the management to review all the planned orders before they are actually released.

The material items should be allocated to individual orders according to the priorities associated with them. Material shortages are often detected at the time of order release in spite of all the detailed planning that might have been carried out. Outside suppliers may not have delivered the items on time. Alternatively the planned materials may have been used up for a rush order.

The production controllers should be able to interact with the system to determine which raw materials or components are holding up the release of an order, thereby eliminating the need for physical staging. This paper staging of items required for an order is critically dependent on the accuracy of inventory records since with inaccurate records the shortages will not be revealed until the order has been released. These shortage reports can then be used by production or purchase departments for expediting the availability of required items.

In the event of shortage of required items, the normal procedure should be to reschedule the planned order release to a later date when short items will actually be available. However the system should be flexible and allow the management to override this automatic rescheduling, i.e. if necessary the management should be able to force release an order. Such forced release of an order can be very useful in the following circumstances.

(a) If the shortage items are expected to be available on the date on which the first operation will commence, then time can be saved by starting the process of preparing necessary shop documentation and obtaining required engineering drawings. For example, if items A,B,C, and D are required for the start of production, and the latter two items are not currently available, then the order can still be force released on the planned date. Fig 11.2 illustrates the forced release of an order under these conditions.

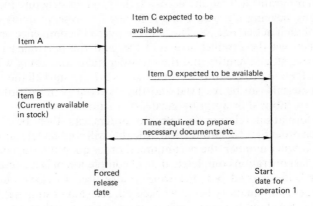

Fig. 11.2 Forced release of a production order

(b) In some circumstances it might not be necessary to delay the order release because the production can proceed without the availability of some items which are only temporarily out of stock.
(c) Where it is possible to use an alternative material the production planner can authorize this and release the order for production.

Thus by the forced release of an order the planned production schedule is not affected due to the temporary shortage of some items.

The system should include the facility to modify order release dates for planned orders. This would be required if the actual demand is higher or lower than the expected demand on which planned orders were based. It might also be necessary to alter the order release date, calculated during the requirements planning phase, to balance the load profile.

Any alterations to order release dates also affect lower level items. Best results can be achieved if the system includes a facility to determine rapidly the net effect on lower level items without having to process the complete master production schedule. It should be possible to by-pass the normal planning procedures for the release of rush orders manually input to the system so that the time interval between the receipt of an order and the commencement of the first production operation on it is minimized.

Once the order release is confirmed the system should be able to produce automatically all the shop order identification documentation, manufacturing routes, feedback documents, material pick up lists, and requisition slips.

Order Release Programs

Each planned order, stored on the planned order file, has an associated release date. By obtaining access to this data all the orders due for release between two dates which specify the order release horizon can be listed on a temporary file. This list can then be used to allocate and reserve materials for different orders taking into account the priorities associated with individual orders. The material allocation programs will require access to the product structure file, part master file and any pegging file which might be used to reserve items for orders in advance of their actual release. The items required to commence production can be identified via the product structure file, and the part master file lists the available free stock quantity and the on-order quantities along with associated due dates. If enough quantities of components exist to start all the orders due for release, materials can be reserved and the status of planned orders altered to 'released' (without shortages) by means of a suitable code.

The information relating to released orders can then be stored on the production order file. The production order file will contain relevant data such as production order number, the part number, order quantity, start date, due date, lead time, last operation completed, date of completion of last operation, current operation being carried out, the work centre used to carry out the current operation, current quantity being worked on, relevant costing data, the pointer to the next operation, the next work centre number, and the address of the next production order for the same item.

If, as is often the case, component and material shortages exist then some of the orders will have to be held back. The status of these orders can be changed to unreleased (due to material shortages). This information can then be displayed on the screen of a v.d.u. or printed out. At this stage the production planner can interact with the system to check the feasibility of forced release of orders held back. This interaction can best be carried out in an on-line conversational mode. The possible use of alternative materials, the anticipated receipt of items from outside suppliers, or the expected completion date of production orders can be examined before the forced release of an order. In that case the status of the order should be changed to released (with shortages). The printed list of shortage items can be used as an exception report to expedite their availability. This shortage information should also be linked to the records of the relevant unreleased (due to material shortages) or released (with shortages) orders so that when these items are received, the orders being held up can be identified and action taken to progress them to the next stage.

Preparation of Shop Documentation

Following the confirmation of order release the system can then produce the documents required to accompany the order. A shop packet typically consists of the engineering drawings of items to be manufactured, a routing sheet showing the manufacturing route taken by the job, material pick up list and feedback documents. In addition paperwork is required to show, for each manufacturing operation, the work centre in which the operation takes place, the materials and tools needed, and the detailed manufacturing instructions.

The production order identification document, issued to the the foreman, is essentially an authorization to manufacture the item and contains information such as production order number, part number of the item being produced, engineering drawing number, manufacturing cycle time, start date and finish date. In the case of pegged production items the order identification document will also show the customer name and order number. The shop order identification document, if in the form of a punched card, can also be used to report the completion of operations on an order. This punched card and an operator identification badge can be input to a shop-floor data collection terminal, and the variable data entered via a keyboard.

The operations document usually shows the production order number, operation number, operation sequence number, its description, the work centre at which it is performed, scheduled start date, and scheduled completion date. It may take the form of a printed form. Alternatively it may be a punched card which can also be used as a turnaround document. Variable data such as quantity produced and scrapped, the actual time of completion of an operation, and the operator check number can be input to a shop-floor data collection terminal. Alternatively these variable data are written on the turnaround document and used later to update the production order file.

The routing sheet is another document which accompanies the order through all stages of the production process. Typically the routing sheet lists, for all production operations, their sequence numbers, operation number, scheduled start and completion dates, work centre at which an operation is to be performed, and their detailed descriptions. This last item of information ensures that the operations are performed on the shop-floor according to the instructions prepared by the production engineering department.

Before manufacturing can start on a batch of work all the materials and components required must be collected. The details of items required are shown on another document, commonly referred to as the kit marshalling list, which forms part of the production order documentation. A copy of this document is also issued to the stores personnel so that the availability of required items can be assured.

Material requisition documents containing details of individual components to be withdrawn from the store can be in the form of printed forms or interpreted punched cards; the latter can be input to the computer system on completion of the transaction. The material requisition practices vary from one company to another. In some companies all the materials and components required to complete an order are issued at the commencement of the first operation. An alternative and better procedure is to issue initially materials required for the completion of the first operation only. By simulating the completion of the previous operation and the start of next, the time of issue of the materials required for subsequent operations can be determined. As a result the quantity of material on the shop-floor, at any given time, is diminished with a consequent reduction in material handling costs.

Shop Documentation Creation Programs

The input to the shop documentation creation programs consists of orders whose status has been altered from 'planned' to 'released' for production (with or without shortages). The detailed data for each production order includes part number, quantity required, manufacturing start date, and manufacturing due date. The production order identification document or ticket can be prepared by obtaining access to this production order record, and the relevant fields in the part master file from which data items such as the engineering drawing number and manufacturing cycle time can be retrieved.

The routing sheet can be produced by obtaining access to the disc address, stored in the part master file, at which details of the first production operation on the item are stored in the routing file.

The details of subsequent production operations can be retrieved by following the routing file chain which links, in a sequential manner, all the operations to be performed on the required item. Access to required work centre data is obtained via the routing file which contains a pointer to the record of the work centre at which a particular operation is carried out. The required feedback documents can also be prepared as part of this program.

The kit marshalling (or material pick up) list and the material requisition documents are prepared by obtaining access to the disc address of the product structure file at which relevant bill of materials data, for the required item, are stored. Detailed information about individual lower level items can be retrieved by following the pointers from the product structure file to the part master file.

Sometimes the planned material is inadequate for continuing the progress of a production order, as for example in the case of excess scrap on a production order. With a batch processed system the generation of a new material requisition can take considerable time and it might be necessary to hold up the order. The issue of material against handwritten documents which are not posted to the computer often leads to a corruption of the inventory record. In such circumstances it is desirable that the system should include a facility whereby the shop foreman can input directly, via a terminal, the request for a new material requisition. If the item required is available in stock, then a new requisition can be created immediately. Otherwise the terminal operator can be informed about the date on which the required item is expected to be available.

Fig. 11.3 illustrates the flow of information in an overall order release system.

Operations Scheduling

The manufacture of a product involves the performance of a number of operations, in a pre-defined sequence, at a number of work centres. Since it is usually necessary to ensure maximum utilization of machine and manpower capacities, it is seldom possible to start work on the job immediately after it

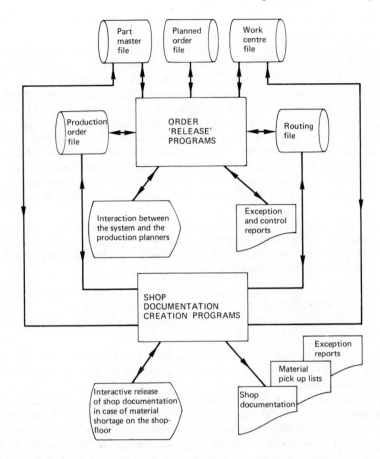

Fig. 11.3 Information flow in an order release system

arrives at the work centre. As a result most jobs have to wait in a queue before work can start on them. The amount of time spent in the queue varies according to the product mix and volume of jobs. Sometimes it might be possible to start work on a batch immediately after it arrives in the work centre. At other times there might be considerable delay. This uncertainty about the waiting time makes it difficult to simulate accurately, for a long period ahead, the sequence in which operations are carried out on different machines. It is also difficult to predict machine breakdown. When a machine breaks down all the operations scheduled to be carried out on this particular machine are delayed. This also affects the sequence of operations in other work centres. Therefore any attempt to establish a daily working list too far in advance results in unrealistic schedules since the activities on the shop-floor will, in the meantime, have altered the priorities allocated to various orders.

The major objective of operations scheduling process is to prepare a realistic production schedule; only the efficient execution of such a work list can ensure that orders are completed on time and the work-in-process inventory is minimized. Another objective of an operations scheduling system is to attempt to minimize the idle machine and manpower time.

Operations scheduling primarily involves the establishment of expected start and completion times of a job. The main factor which contributes to the complexity of the operations scheduling is the very large number of batches of work moving through the production plant at any given time. All these batches of work compete for a limited amount of available production facilities. The start of an operation on a particular batch is dependent upon the completion of the previous operation, and as soon as the current operation is completed, the same batch of work starts competing for the facilities required to carry out the next operation. As a result the choice of one from a number of jobs, requiring the same production facilities, is usually dependent on the priorities associated with individual jobs. The priority allocated is dynamic and varies from one day to next.

Operations scheduling can be carried out by simulating the load queues, machine set-up and run times, and transportation times, at various work centres, and then loading the work centre to its actual capacity. The following criteria can be used to load an operation on a given work centre.

(a) It is the first operation to be performed on the particular batch of work.
(b) All previous operations on the batch have been completed.
(c) All previous operations have been loaded during the running of the operations scheduling program.

If shop load levelling has not been carried out then the total volume of work on which operations have to be performed, in a given work centre, might be more than the available capacity.

The limiting factor is the available machine capacity or more usually the availabilty of operators. If shop load levelling has been carried out, and the possible use of alternative manufacturing facilities considered, then the work can be loaded onto the selected alternative work centres. In the absence of such alternatives it will be necessary to load the job on the work centre specified in the routing file. At this stage it becomes necessary to calculate a priority for each order or operation so that high priority operations are loaded before low priority operations.

Rush Job Scheduling

If the latest start date, calculated during infinite capacity backward loading, falls before the current date then the rescheduling of the order from the current date results in a delayed delivery. It is often necessary to attempt to complete the order by the due date so that penalty clauses, which might become effective in the event of delayed delivery, are avoided. This can be achieved by reducing the queue and transportation times. An alternative is to reduce the manufacturing cycle time by splitting and overlapping the batch for performing certain specified operations.

Order Splitting

When an order quantity is split the objective is to reduce the overall cycle time by simultaneously performing the work on more than one machine. As a result the total run time is divided by the number of machines which are in use while the set-up time is multiplied by the same number of machines. Each sub-batch of work requires separate tools and set-up. The split usually applies only to one operation, and the sub-batches are re-consolidated into one before the start of the following operations.

The result of splitting a batch of work into two smaller batches is shown in Fig. 11.4. The run time is halved, while the set-up time is doubled thereby resulting in higher costs.

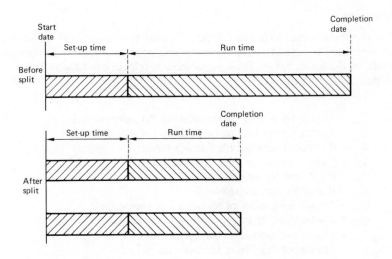

Fig. 11.4 Graphic illustration of the effect of splitting an order

The split can be economic or forced. Economic split is used after taking into account the economic factors. By pre-specifying a maximum value of the split factor, defined below, the management can ensure that an order is not split if it will result in the set-up hours exceeding the run time hours. A split factor can be calculated using the following relationship.

$$\text{Split factor} = \frac{\text{Set-up time}}{\text{Set-up time} + \text{Run time}}$$

The economic split factor may be calculated or specified by the management.

Under certain circumstances the management might consider it necessary to split an order irrespective of whether or not the split is economic. In such cases the number of sub-batches is specified by the management.

It is also necessary to consider, in both cases, the technical constraints such as the number of men or machines available at the required work centre, and number of tools required. For economic split the management can also specify the maximum number of sub-batches into which a batch quantity should be split. Thus the number of sub-batches (n) into which a batch quantity is economically split is the highest value which satisfies the following relationships:

$n \le$ Number of available tools.

$n \le$ Number of available operators or machines.

$$n = \frac{\text{Run time}}{\text{Split factor} \times \text{Set-up time}}.$$

$n \le$ Maximum number of sub-batches.

Overlapping

An operation is said to be overlapped if a partial quantity is sent ahead to the next operation, referred to as the overlapping operation, before the completion of work on the entire batch. This overlapping results in a reduction in the overall lead time as shown in Fig. 11.5.

The variables in Fig. 11.5 are as follows:

$Q_{p1} =$ Queue time prior to processing for operation 1
$Q_{p2} =$ Queue time prior to processing for operation 2
$S_{t1} =$ Estimated set-up time for operation 1
$S_{t2} =$ Estimated set-up time for operation 2
$R_{t1} =$ Run time for operation 1
$R_{t2} =$ Run time for operation 2
$Q_{a1} =$ Queue time after processing for operation 1
$Q_{a2} =$ Queue time after processing for operation 2
$T_{t1} =$ Transportation time for operation 1
$T_{t2} =$ Transportation time for operation 2
$T_{tr} =$ Reduced transportation time

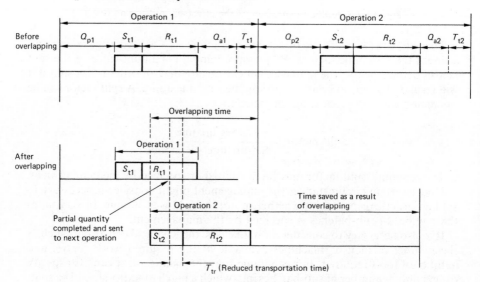

Fig. 11.5 Graphic illustration of the effect of overlapping two operations

Since the batch is being split in order to ensure that it may be worked upon, it is reasonable to assume that the queueing time at the next operation will be zero, and the transportation time can also be reduced by a high percentage.

Certain operations may not be overlapped. For example heat treatment and similar processes which are carried out in a batch should not be overlapped.

Overlapping can be economic or forced. A number of constraints have to be taken into account for economic overlapping. The overlapping operation, once started, should not be interrupted. This means that the quantity sent ahead must be such that the next sub-batch sent ahead from the overlapped operation is received before the completion of the overlapping operation on the previous quantity sent ahead. Therefore a minimum 'send ahead' quantity has to be specified, and worked upon at the overlapped operation.

Another constraint is that the set-up of the overlapping operation should not commence before the start of overlapped operation set-up. These constraints can be mathematically represented as;

$$T_{01} = R_{t1} (Q - Q_s) + S_{ts} - T_{tr}$$
$$T_{02} = R_{t2} (Q - Q_s) + S_{t2} - T_{tr}$$
$$T_{03} = R_{t1} (Q) + S_{t1}$$

where

T_{01} = Overlap time, constraint 1
T_{02} = Overlap time, constraint 2
T_{03} = Overlap time, constraint 3
R_{t1} = Run time for operation 1
R_{t2} = Run time for operation 2
S_{t1} = Set-up time for operation 1
S_{t2} = Set-up time for operation 2
T_{tr} = Reduced transportation time
Q = Order quantity
Q_s = Minimum quantity sent ahead

The selected overlap time is one which satisfies all three constraints and is the minimum of the three quantities calculated above.

Overlap time $T_0 = \text{Min} (T_{01}, T_{02}, T_{03})$

The feasibility of economic overlapping can be checked by comparing the calculated overlap time against the minimum overlap time specified by the management. If the calculated overlap time is greater than the management specified minimum overlap time, then the new starting date for the overlapping operation is calculated according to the following relationship;

$$\binom{\text{New start date}}{\text{for operation 2}} = \binom{\text{Old start date}}{\text{for operation 2}} - (T_0 + T_{t1} + Q_{p2})$$

Where Q_{p2} = queue time prior to processing for the second operation.

Under certain circumstances the management might wish to carry out overlapping irrespective of whether or not it is economical. In such cases the management specifies the minimum send ahead quantity and the time at which the set-up for the second operation should start.

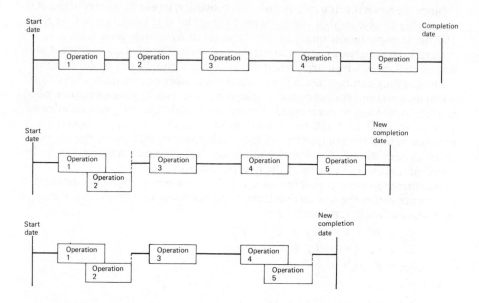

Fig. 11.6 Graphic illustration of the effect of overlapping a number of operations

If the time gained by the overlapping of two operations is inadequate, then it might be necessary to overlap further operations. Fig. 11.6 shows the times saved by overlapping operations 1, 2, and 4, 5 respectively. It is assumed that operation 3 cannot be overlapped due to the nature of the process involved.

Priority Rules

The priorities associated with individual operations can be developed in a number of ways. It is desirable that the highest priority should be associated with operations which have been partially completed, i.e. operations being completed.

In order to achieve the maximum benefits from overlapping, it is also desirable that a job with an operation which overlaps a previous operation is loaded as soon as possible. Similarly an order which was split and then reconsolidated for the next operation should also be loaded as soon as possible so that the overall objective of reducing manufacturing cycle time is achieved.

Consecutive operations which require the same set-up should also have high priority.

An operation relating to an order for which the latest start date was on or before the current date can also be accorded high priority.

The management might decide to allocate a high priority to jobs on which a number of succesive operations are performed in the same work centre.

Operations with relatively short processing times can also be allocated high priority. Thus, an order with an operation which has the shortest processing time will have priority over other orders. As a consequence the manufacturing cycle is reduced and the resource utilization improved.

The relative priorities associated with any of these conditions will vary from one company to another according to their specific requirements. In addition to these special priorities, internal order priorities can be calculated taking into account the following factors:

(a) Operations processing time.
(b) Total number of remaining operations.
(c) Time available to complete the job.

$$\text{Internal priority} = \frac{\text{Due date} - \text{Current date} - \left(\begin{array}{l}\text{Processing time for operations} \\ \text{which have to be completed}\end{array}\right)}{\text{Number of operations which have to be completed}}$$

Thus, the internal priority to be associated with individual orders is calculated again every time a new system run is carried out.

The company management can achieve considerable flexibility by ensuring that the system accords a high overall priority to an order which has been allocated high external priority by the management. This facility is particularly useful when management quotes an earlier than normal delivery date in anticipation of a large follow-up order.

Operations Scheduling Programs

Over a specified future period of time the work of a manufacturing plant can be simulated by retrieving the records of released orders, various operations which must be performed during this period, and the work centre data. A matrix can be used to represent the load on individual work centres in specified future periods. As the capacity is used up, the relevant fields in the matrix can be appropriately incremented to reflect the increased load.

In order to ensure that orders with high priorities are loaded first, all the orders waiting to be processed in a particular work centre can be collected in a file and then sorted according to their priorities.

The jobs currently in process should be loaded first since, under normal circumstances, it is usually uneconomic to interrupt the work being carried out. A parameter can be used to override, if necessary, the continuation of work on the current batch. The remaining jobs can then be loaded according to their priorities. Once an operation is loaded it is removed from the temporary file. Operation start and completion times estimated during the loading process can be stored and printed out on the job sequence list for the particular work centre.

By taking into account the queue time after processing, and the transportation time for an order completed in a work centre, the time of arrival of this order at the following work centre is estimated. The queue time after processing and transportation time might be the one associated with the given operation. If such time is not specified then an average value for the work centre can be used. The job can then be entered into the queue for the work centre at which the next operation will be performed. This process is repeated for all the jobs on which operations are to be carried out during the operations scheduling time period, for instance a shift, day, or week.

If adequate capacity is not available in the work centre in which a given operation is normally performed, then it can be loaded onto an alternative work centre, if one exists. The work centre record should contain a pointer to the alternative work centre. Under normal circumstances such loading should be carried out only if it would result in the job reaching the next work centre earlier than if loaded on the usual work centre. However, under special circumstances, for example to improve the utilization of work centres working below their normal capacity, an alternative work centre might be specified even if it does not satisfy the criterion mentioned above. A suitable parameter can be used to override the normal condition. The actual loading process is very similar to the one specified above i.e. the loading is carried out taking into account the priorities associated with individual operations.

Once all the operations have been loaded, a list of the sequence of operations to be performed at individual work centres along with a load analysis report is prepared. For each job listed on the work sequence list, additional information such as set-up time, run time, operation number, priority associated with the operation, the batch size, previous work centre, next work centre, expected arrival time, expected start time, expected operation completion time should be included to help the foreman carry out his work in an effective manner. Any special action such as order splitting and overlapping should be highlighted. Where forced splitting or overlapping has been carried out the system should still

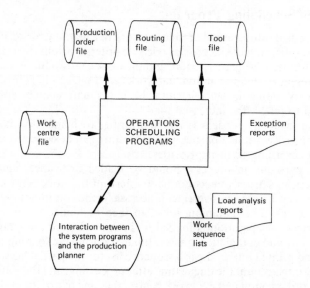

Fig. 11.7 Information flow in a typical operations scheduling system

perform the economic calculations and show the resulting adverse economic effect. Similarly delayed orders, i.e. orders not scheduled due to lack of capacity, should also be highlighted to bring them to the attention of the management.

The effect of altering the priorities associated with some of the late orders, and/or reducing the queue and transportation times by a specified percentage, on the expected completion time and the work centre load can be estimated. Fig. 11.7 illustrates the flow of information in a typical operations scheduling system.

The simulation of future operations can be carried one stage further. The production manager can then use the resulting load information for planning the required overtime working. Such simulations become practical tools when used as an integral part of an on-line interactive production planning and control system.

Summary

A computer-based order release system is preferred to its manual counterpart due to the following distinct advantages which can be achieved.

The need for manual staging of all the materials required to start work on an order is eliminated. Reports relating to material and tool shortages can easily be prepared by interacting with the system. The possibility of forced release of an order, and its impact on other orders can quickly be examined. Orders which cannot be released due to shortages etc. are highlighted by means of exception reports. The management can review the order a short time prior to its actual release and make any alterations to the order before the necessary shop documentation is prepared.

The operations scheduling process, which cannot be carried out in the absence of a computer, make it possible to simulate, in the short term, all the operations carried out on the shop-floor by taking into account the actual set-up and run times, as well as the queue and transportation times. The priorities associated with individual operations are fully taken into account during loading to available capacity. This makes it possible to prepare a realistic work schedule for the following day or shift. If it is found that the work lists do not take into account the current conditions on the shop-floor, due to for example the prolonged breakdown of machines or the absence of a number of key operators, the system programs can be rerun and a new work schedule prepared.

12
Work-in-Process Control Systems

Introduction

The work-in-process control system forms the interface between the planned operations and actual events which take place on the shop-floor following the release of the order. Many shop-floor problems are caused by inadequate planning. In the absence of proper planning precedures, the production orders are often split, and overlapping of operations becomes a common occurrence. Planning, no matter however detailed, does not always represent the shop-floor activities. The sequence in which operations are actually carried out is not always according to the plans. If it were possible to carry out all the operations as planned, then it would not be necessary to use a work-in-process control system. Such conditions are rarely, if ever, encountered in actual practice. The shop-floor environment is a dynamic system which is affected by a large number of random events such as the breakdown of machines, absence of key workers, non-availability of materials or tools. For efficient management of resources the production planners and managers have to be aware of the up to date status of the work-in-process (WIP). As a consequence it is necessary to monitor the actual condition of the activities which have been planned. The management can then take any necessary corrective action. If such corrective action cannot be taken, due to capacity constraints etc., then it will be necessary to alter the production plans. The management can also ensure that the work is making planned progress, and that the actual utilization of machines, manpower, and materials is not significantly different from their planned utilization levels. Also if the quantity passed on from one stage to the next falls below the finally required quantity, another batch of work can quickly be released to the shop-floor so that the complete order can be delivered on the due date. Similarly if the manufacturing process results in lower than expected yield, the production engineering department can be asked to investigate the process and attempt to rectify any problems which might exist.

Due to the large volume of data involved and the dynamic nature of the shop-floor, it would be difficult, if not impossible, to maintain adequate control over the work-in-process, without employing a number of progress chasers and clerks. In addition, it is not feasible to assess accurately, at any time, the up to date status of the current shop-floor activity. Such detailed information about individual orders is a pre-requisite for quickly making the large number of decisions which the production controller is expected to make during the course of a day. Most manual WIP control systems are inadequate, and the flow of information is often delayed with the result that the decisions are made in the absence of relevant information.

A computer-based work-in-progress control system can be used to keep accurate track of the very large volume of data involved and quickly highlight, by means of exception reports, events which require management attention. With a system in which the relevant summary information is separated from irrelevant details, the management can quickly respond to the changed conditions and breakdowns which take place on the shop-floor. The preparation of a realistic work sequence list, during the running of the operations scheduling program, is dependent on the availability of information relating to the current state of the shop-floor activities. Inadequate or inaccurate data will only result in the creation of a schedule which cannot be used on the shop-floor.

An on-line real-time system used for this application is particularly beneficial. Special purpose shop-floor data collection terminals can be used to report the start and completion of operations on a batch of work. Such terminals are also useful for recording the times of arrival and departure of company employees. As a result the system can monitor the manpower available in different departments. This information can then be compared to the planned or required manpower levels, and, if necessary, some of the work schedules altered to take account of the absence of key workers. Some of the machines, on which manufacturing operations are carried out, can be connected to the computer, and the relevant data collected automatically.

Features of Work-in-Process Control Systems

An effective and fully implemented work-in-process control system should include provision for the feedback of the following data from the shop-floor:

(a) Quantities produced
(b) Quantities scrapped
(c) Material usage
(d) Start of an operation
(e) Completion of an operation/order
(f) Time spent working on an operation/order
(g) Machine breakdown time
(h) Operator waiting time
(i) Other exception conditions such as changes to planned orders, deletion of an operation.

Wherever possible this data should be collected and input to the computer at source and immediately validated so that the production order file is always up to date. The details of shop-floor data collected will vary from one organization to another. Similarly the actual method used to collect this data also varies. The completion of an operation might be reported by returning a pre-punched card along with a record of the actual quantity completed if it differs from the planned quantity. A majority of data relating to the production orders are fixed and can be input to the system by means of special data collection terminals using a pre-punched card. The variable data such as quantity completed or scrapped is entered on a keyboard attached to these terminals. In some cases it might be preferable to vet the shop-floor data manually before it is used to update the computer files. In such systems the shop-floor data are input by an experienced v.d.u. terminal operator familiar with the combinations in which the various data items can occur. The input data are immediately validated by data validation programs, the errors encountered displayed on the terminal screen, and the operator instructed to input correct data. This detailed work-in-process data can then be used for preparing the work schedules for the following shift, day or week.

In systems in which operations scheduling is carried out on a very frequent basis, for example daily or for each shift, it would be possible to reduce the amount of shop-floor data which is reported back to the system. At the end of each shift/day the departmental foreman can check his daily job list and determine the operations which have not been completed. The data relating to the operations not completed are then input to the system. Otherwise it may be assumed that the work has been completed as planned. Thus it is necessary to report WIP data only in the event of exceptions whereby deviations from the plan are reported.

Another alternative is to report the start of each operation and assume that, unless otherwise reported, it will be completed.

The management of some companies, for example electronics industries, engaged in the manufacture of products involving very short cycle times might decide that it is uneconomic to report the start or completion of all the individual operations. In such cases it would be necessary to identify certain important operations, for example the ones resulting in a high percentage of scrapped items. The data relating to the completion of these key operations, along with the quantities passed on to the next stage are input to the system. If the quantity passed on to next stage falls below the required quantity an exception report can be generated so that the management is able to take the necessary corrective action, i.e. arrange for the release of a new batch of work. Similarly, if an operation completion report is not input to the system by the due date for the identified key operation, an exception report should automatically be produced by the system to expedite the progress of the order. Otherwise it can be assumed that the order is satisfactorily progressing from one operation to the next. In such cases it is only necessary to prepare turnaround cards for these key operations. The amount of WIP data, which has to be processed by the system, is also reduced. It is highly desirable that the first production operation is always identified as a key operation, so that the system is aware that the work on the batch or order has actually commenced.

The shop-floor data forms the basis of a number of detailed reports which can be of considerable help to the management. These reports could, for example, show:

(a) Details of critical orders behind schedule so that the management can take action.
(b) The value of the current work-in-process.
(c) The orders completed.
(d) The status of batches currently loaded onto the shop-floor.
(e) A variety of performance analysis indices relating to the level of work centre utilization, individual operator, average operator, departmental and factory performance.

The work-in-process data is also useful for performing bonus calculations in the case of companies which operate piece work or other bonus schemes.

'Operation complete' type of transactions, for released orders, can be used for preparing reports relating to the value of the work-in-process, the status of orders, performance analysis, as well as any exception conditions. These data are also useful for calculating actual costs, which could show appreciable variations from planned costs if a work centre other than the normal is used for performing the operation. Similarly this data can be used for updating the smoothed average queue time and total time required to complete the operation.

Based on the reports of 'operation complete' transactions, and by subtracting the operation lead time from the total lead time, the remaining order completion time may be calculated. By comparing the actual shop-floor activities with the planned activities a report showing the difference between the two can be prepared for management action. This information can then be used during the next run of the operations scheduling system. For the preparation of a realistic daily working schedule, it will be necessary to take into account operations partially completed at the end of a shift or a day. The report for such a transaction should indicate the partial quantity which has been completed and the time spent working on that quantity. Since a partial quantity has been completed, the remaining operation time can be calculated by subtracting the queue time before processing, the set-up time, and the run time for the reported quantity, from the total allowed time.

The report relating to the completion of an operation usually implies that all the previous planned operations have been completed. In general, operations in a shop are performed in the sequence in which they are planned. However, in practice, it is often the case that it is better to perform a later operation earlier on in the sequence; for example if a machine is already set it might be more economical to carry out operation number fifty before number forty. In such cases it is essential that the operation complete/started transaction indicates that the work is running out of sequence. Otherwise it will be assumed that all the previous operations have been completed and the operations missed out will not be taken into account in subsequent scheduling, thereby resulting in the due date not being met. If the manufacturing route has been altered then this change should be reflected in the temporary records relating to that order. In an on-line real-time system this can easily be achieved by entering the changed manufacturing route, via a user terminal, to the computer system. The new or revised paperwork can be generated simply thereby reducing the effort required to implement any required changes.

On completion of the final operation, the data can be input to the system either in the form of 'operation, complete' or 'order complete' transaction and the order closed.

The system should include provision for the report of the use of unplanned labour so that any additional work, for example on scrapped quantities, can be reported and taken into account in costing calculations and in the preparation of reports relating to the value of work-in-process, the status of orders and performance analysis.

Data relating to the material actually issued against material requisitions, as well as further material issued in the case of scrapped items, should be input to the system and used to calculate the actual product cost.

Considerable flexibility can be built into the system by first entering a code number, which indicates the type of transaction involved, followed by further data such as operator number, operation number, quantity completed, time spent on the job and work in hand.

Tool Control

An important consideration in the preparation of work sequence lists, using the operations scheduling system, is the availability of required tools since no work is performed in their absence. As a consequence it becomes necessary to plan for the use of these tools. It is also desirable to monitor the use of tools so that necessary preventive maintenance, based on their actual usage, can be carried out. The time over which a tool is used may be determined in two different ways. In the first case, every time a tool is used, the relevant usage data can be input to the system. Alternatively it can be assumed, within the system, that a scheduled tool has been used unless an exception report regarding the non-availability of a tool is input to the system. Based on these tool non-availability transactions, an exception report for appropriate management action, such as making provision for extra tools can be prepared. It would also be necessary to input tool usage information specifically, if the same tool identification is used for a number of similar tools. This actual tool usage information is then used for preventive maintenance in accordance with a code associated with the tool.

Monitoring of Work Queues

As part of the simulation of activities on the shop-floor, it is necessary to make an estimate of the average queue times before and after processing. It is desirable that this data should be updated at regular intervals. This may be achieved by monitoring, at pre-defined or random times, the actual number of hours of work waiting at the work centre. The latest data is smoothed with old queue times, using exponential smoothing techniques and updated values of queue times calculated.

Assignment of Jobs

In many companies jobs are assigned to individual operators by the foreman. The availability of an on-line real-time computer system makes it possible for him to review the up to date status of all the jobs which are to be processed in a particular work centre. He should be able to take account of other jobs which are expected to arrive in the work centre during the course of the next few hours. He can then attempt to minimize the total set-up and manufacturing cycle times by combining jobs with similar set-ups. In addition he can use his intuition and knowledge to select the jobs which require close tolerances and the machines on which these jobs should be processed. Similarly he can take into account the skills of individual operators. It would be difficult to consider all these factors during automatic loading carried out as part of the operations scheduling process. Where this information about the assignment of jobs to individual operators is recorded on the production database, the operators can use a terminal to enquire about the next job on which they should work; otherwise this information can be printed out along with other relevant data such as the need to split or overlap operations.

Validation of Shop-Floor Transactions

A large volume of shop-floor data is used to update the production database. To ensure accuracy and integrity of the database all this data should be checked carefully for accuracy and credibility. Some of these checks can be performed against the data held in the production database. For example, the operation for which data is currently being entered can be checked against the last reported operation. Similarly the data relating to the quantity passed on from one stage to the next can be checked to ensure that it does not exceed the previously reported quantity. As errors are detected the operators can be asked to correct them. If the currently entered data are correct and the error occurred at a previous stage, then the foreman or a person with higher authority should be able to input such apparently erroneous data. In some cases it might be necessary to include a facility to cancel a previous wrongly entered transaction by using a different code. It is essential that only a few people should be allowed to override the errors reported by the system. The above facilities can only be used with an on-line data collection system. In the case of batch processed systems, the erroneous data are rejected and listed on an exception report, and the users asked to input correct data during the next run of the system. However, this procedure does mean that operations scheduling is carried out on the basis of data which does not represent the true shop-floor activity state.

Expediting of Jobs

It is often necessary to expedite the processing of some critical orders which are either late or are urgently required. In manual systems the expediting involves physically shifting items to the next work centre and asking the supervisor to ensure that the particular job is worked upon next. This function is usually performed by a number of progress chasers. The system often fails because the management is unaware of the effect of this expediting on other orders. Also managers, foremen, and progress chasers have different priorities.

In computerized operations scheduling and work-in-process control systems, an order can be expedited by altering the priority associated with the particular order. The higher the priority, the further up the order will appear in the work sequence list. To ensure that this facility is not misused only senior managers should be allowed to alter this priority, otherwise the priority system will simply break down. With such a computer-based expediting system it is no longer necessary to negotiate with a number of people in affected departments.

Material Handling Control

Material handling costs represent a significant proportion of the total manufacturing costs. Labourers are employed in some companies to move items from one area to another. Often, however, the production operators have to spend time collecting required materials and tools from other areas. As a result the production efficiency suffers. It is desirable to use a system such that the need for operators to collect items is minimized. In an on-line system this can be easily achieved. Once the foremen have assigned orders to different machines, all the items required to start an operation can be identified and the relevant personnel instructed to move them near the appropriate machine rather than to a general area. The priorities associated with individual orders are also taken into account in that items associated with high priority orders will appear higher up on the list and can be moved before low priority orders. New move lists may be prepared at hourly or other suitable intervals. These move lists should identify the items to be moved, their current location and the location of the machine to which they should be moved so that the time required to search for these items is minimized. With the use of such a system the time spent by the production operators in moving items is considerably reduced.

Quality Control

It is the function of the quality control department to ensure that all products meet the quality tolerances specified by the company management. Many companies use statistical quality control procedures while others test every single item. It is desirable that, irrespective of the quality control procedure used, the items should be tested as soon as possible after the completion of the relevant operation. If it is found that the quality is below standard due to some process faults, then the production of further items can be stopped thereby preventing further scrap.

The inspection operations can be planned using procedures similar to ones used for production scheduling. The standard inspection times may be developed using historical data. In some cases the inspection operation and the time required may not be separately identified, and the operation may be performed immediately after the completion of the production operation. The inspection time is allowed for in the total production time.

Where the inspection operation is performed separately, it is necessary to post the resulting data to the database. Procedures similar to the ones used for reporting completion of production operations are employed for this purpose. The only additional data, required for reporting the completion of an inspection operation, relates to the reason for scrapping some or all of the production items. The reasons for scrapping items can be specified in terms of appropriate codes.

Computer-based automatic testing procedures are becoming increasingly popular in manufacturing industries. The testing process is directly controlled by the computer, resulting in an improvement in accuracy as well as reduced costs. For complex products, manufactured in large numbers, the computer can initiate the appropriate sequence of tests, and decide on the suitability of the item. Some of the currently available computer-based testing systems include facilities to check that the testing system itself is performing satisfactorily. This is a desirable feature due to the fact that the test equipment is often more complex than the component under test. The data collected during the testing of items can be used to update the production database automatically by employing a suitable interface program.

Direct Control of Production Operations

As the volume and complexity of products required to satisfy the customer demand continues to increase, the managers of manufacturing companies are finding it beneficial to invest in automatic machinery. The cost of computer hardware, in particular mini- and micro-computers, has been declining steadily. As a result the use of such computers for controlling manufacturing operations has become a practical proposition. Computer numerical control (CNC) and Direct numerical control (DNC) systems are being used in preference to 'hard wired' numerical control machines. The use of such equipment not only results in higher production efficiency but also makes it possible to collect directly the work-in-process data by connecting the computer to the production machinery. Transducers in the form of sensing and counting devices are connected to the computer and used to keep track of conditions such as the status of the machine, i.e. whether it is idle, being set up, or in production run, the feed rate, number of items produced. The changes in production rate, in particular significant changes, can be highlighted for management attention. The other data, for example, the order number, operator identification number, etc., can be input to the system by means of data entry terminals attached to the production machine and also connected to the computer. This automatic data collection is less time consuming and reduces the possibility of errors. In addition one of the frequently encountered problems, whereby the operators do not report the completion of operations on time in an attempt to get an increased bonus for the next week, is eliminated.

While a majority of the currently used production and testing equipment cannot be used for automatic data collection without expensive modifications, it is nevertheless necessary that the possible future use of such equipment and the need for interface programs should be taken into account fully in the development of an integrated production system. Also, the management should take account of the WIP data collection requirements when considering the purchase of new machinery.

Work-in-Process Control Programs

The flow of information in a typical system is as shown in Fig. 12.1.

The identification number of the production order, for which some transaction data are to be processed, can be used to locate the disc address at which detailed information relating to that order is stored. Codes can be used to distinguish between different types of transactions. The fields affected, in the production order file, will vary according to the transaction being processed. Before the fields are updated, the input data must be validated to check their accuracy and credibility. The processing of input data can best be illustrated by considering the most common type of transaction pertaining to the completion of individual operations. The data relating to the operation carried out and the quantity

G

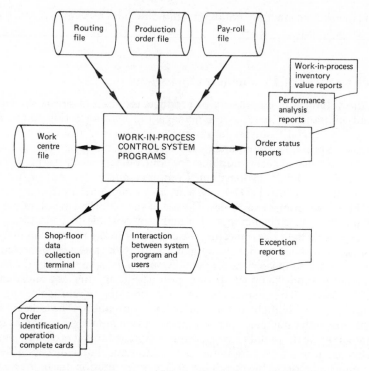

Fig. 12.1 Flow of information in a typical work-in-process control system

completed and passed on to the next stage can be used to update the order record, after the data have been edited. The relevant operation number can be checked against the last recorded operation and the quantity compared to the last reported quantity. In the event of errors an exception report indicating the type of error encountered should be generated, and the user asked to enter correct data. The actual time spent working on the order is used to prepare the necessary performance analysis reports. The data relating to scrapped quantities are required to update the yield information, and if the current yield falls below a specified percentage an exception report can be prepared to draw the attention of the production engineering department. This data can also be used to collect the relevant actual product cost information. The same data are also input to the payroll system used for bonus calculations.

The WIP control system should also include a program which can be used to examine the whole of the current production file at specified times, for example the end of a shift. This will reveal the orders for which expected operation completion transaction have not been posted to the system. Such orders should be highlighted by means of exception reports so that they can be expedited on the shop-floor either by physically moving the items, or by allocating them a higher external priority. These orders will also appear high up on the work sequence lists prepared as part of the operations scheduling system.

The output produced by the system programs is a combination of comprehensive and exception reports. It is desirable that the system should include an option to produce comprehensive reports or exeception reports relating to the status of current WIP. In general, only exception reports should be provided as a matter of routine, and comprehensive reports produced when specifically requested.

Reports may also be prepared to show the differences between the planned and actual activities on the shop-floor. The performance analysis reports can be for individual operators, a group of operators, a work centre, a complete department, or the whole manufacturing plant. In general the details included in routine reports should be kept to a minimum. The work-in-process value reports should show the actual costs incurred to date against the standard costs. The standard cost of the complete production order should also be included so that an estimate of the total actual cost can be made.

In the case of WIP control systems also used for bonus calculations, it might be necessary to prepare a list of transactions reported by each individual employee so that he can check that all the items on which he has worked are being taken into account.

Summary

The use of a computerized work-in-process control system enables the management to co-ordinate closely all the manufacturing operations and the services required to support production. With an on-line WIP control system, the foremen can spend more of their time trying to improve production control rather than handling detailed paperwork. The deviations from planned operations can be highlighted as exception conditions and the management can then intervene. The requirement for detailed minute-to-minute intervention is reduced. Decisions are made by taking into account fully all the relevant facts. With direct computer monitoring and control of production operations the data collection errors are minimized. The accurate work-in-process data can then be used to prepare realistic working schedules for the following shift or day. The number of shop-floor progress chasers required to search and expedite orders is reduced.

13
Inventory Control Systems

Introduction

The efficient management of inventory is one of the major problems faced by managers of industrial companies. The inventory maintained by companies often represents the biggest individual investment in industrial resources, and inventory carrying costs are frequently as high as 30% per annum.

It is impossible to match the demand and supply at all times unless plants with unlimited spare manufacturing capacity are available. In practice this is rarely possible. Inventory is therefore maintained to absorb fluctuations in demand and supply, and improve the level of customer service. High inventory levels imply that a very large amount of capital is tied up. In addition to the unit cost of inventory items, there are also the costs of storing, handling, and insuring the inventory. With slowly moving products it might be a long time before the investment in inventory is recovered. With continuous product development the risk of a product becoming obsolete also increases. Lack of adequate stock can result in the loss of business and customer goodwill. There is traditionally a conflict between different departments of a company. The sales department is interested in keeping a comparatively large amount of inventory so that customer demand can be satisfied, as much as possible, off the shelf to reduce the competition. The financial department wishes to reduce the investment in inventory in an attempt to increase the availability of working capital. The production department, wishing to increase the operational efficiency and to avoid fluctuations in manpower levels, is interested in stabilizing production even if it means having to build up inventory levels. The problem is to strike a balance between too high and too low inventory levels so that the capital invested in inventory can be minimized, while providing a high level of service. In addition to the level of inventory of finished items there are the problems of raw materials and bought out components.

It is a well known fact that a high percentage of inventory investment is usually tied up in a comparatively small percentage of components. The small number of parts responsible for this high level of investment undergo continuous changes over a period of time. Under these conditions a large number of slowly moving items represent high and often inflated inventory values. Conversely there is not sufficient quantity of fast moving items, for which there is heavy demand. This results in frequent stock out conditions. Another commonly encountered problem is an unbalanced supply of associated items, which are usually sold or used together, leading to lost opportunities. The attempt to control the level of capital investment by using the same policy towards all the items stocked by the company does not result in any improvements in the situation. It is necessary to monitor the changes in the pattern of demand relating to various items, and then make decisions about the optimum level of inventory which should be maintained, for these items, at different times. Unfortunately, however, it is difficult to use these techniques in most manufacturing organizations due to the amount of detailed effort required to develop and apply different types of control to various items.

For satisfactory management of the inventory a vital element is the timely flow of information so that relevant decisions are made at the appropriate time. An inventory system is dynamic due to the fact that materials are withdrawn from or added to the stock at random intervals of time. Frequently a quantity of parts, used for production as well as spares, is simultaneously earmarked for use in a machine and for possible sale to an outside customer without any formal records being maintained. This often results in bad feeling because customers are quoted a delivery date based on the ostensible availability of the item whereas, in the meantime, the part is used in the machine building program.

A well designed computer system can resolve these and a number of other problems which are encountered during the formulation and implementation of an inventory control system. Traditionally a number of companies have implemented computer-based inventory control systems since the amount of clerical work required to keep track of a large number of items can be reduced. The only alternative to a comprehensive computerized inventory control system is the employment of a large number of people.

Provided the management is prepared to invest in the required manpower levels a good manual control system can be designed and implemented, but the results are often disappointing due to the fact that the speed at which people can react to changed circumstances is slow compared to the speed of the computer. For example it is a major effort to determine factors such as the stock levels which should be maintained to satisfy a given service level, and economic batch quantities for individual items. These factors, once established, require frequent updating due to changes in the pattern of demand and the prices of individual items. It is also a fact that with manual systems the management is sometimes unable to ensure that the formal system is actually used by all the personnel involved. Often the policy decided by the management is replaced by an informal parallel system. A formal inventory control system which is easily monitored by the management can resolve some of these difficulties.

Even a small reduction in the inventory/turnover ratio, possible by implementing a well conceived computer-based system, can justify the implementation of a computer system which can then also be used for a wider range of applications.

Role of the Computer in Inventory Control Systems

The essential role of the computer in an inventory control system is to help the management exercise adequate control over individual items of inventory by providing it with relevant and up to date information following a detailed analysis of transactions data. In any inventory control system the major effort is spent in processing the large number of transactions which take place at random intervals of time. All these transactions can be validated, processed and pre-digested by the computer and the required summary reports and exception reports produced for management information and action.

In a computerized inventory control system the management of inventory will still be carried out by responsible personnel. The computer is only a tool which can be used to develop detailed policies and implement them by executing suitable computer programs. However, the management will still have to intervene to deal with exception conditions. It is prohibitively expensive to develop a system which can cater for all the exception conditions.

Manual systems cannot provide the detailed analysis required to determine the type of control which should be exercised over individual items, and it is difficult to analyse the past demand and make accurate forecasts of future demand taking into account the trends and seasonal fluctuations which might be present. With a computer these functions can easily be carried out as a matter of routine. This enables the management to exercise different types of control over different items. Also as the pattern of demand changes the policy used in relation to the affected items can easily be modified.

The routine decisions which do not require the individual attention of the management may be automated. However, it is necessary to ensure that facilities for manual intervention are also provided. Any attempt to over-automate the inventory control system is only likely to result in an inflexible system and unpredictable results. The facility for manual intervention is necessary so that if the prevailing conditions are not in line with the assumptions made in the design of the system then the management can easily cope with these changed circumstances.

Computers can also be used to process purchase orders by helping the buyer to select a preferred supplier, and follow up the orders at suitable time intervals. The supplier performance rating can be updated as new deliveries are received, and the information used for selecting suitable suppliers.

In addition, the computer may be used in simulation mode to analyse the effect of changes in the values of system variables. For example, the use of different forecasting and ordering techniques can be investigated, for a number of values of the parameters, before implementing the most appropriate technique. Similarly the impact of any cost changes can readily be tested.

The availability of the computer power as well as the detailed operational and costing data makes it possible to apply simple as well as sophisticated mathematical and operations research techniques.

Let us now consider in more detail the major factors in developing a computerized inventory control system.

Considerations in the Development of Computer-Based Inventory Control Systems

The objectives of an inventory control system, as part of the overall system, do not always remain the same. During periods of cash squeeze it will become necessary to release capital by allowing the stock levels to deplete even if it results in occasional stock outs and lost orders. Similarly it would be necessary to buy in smaller quantities and lose the quantity discounts which might be available. In general, however, it is desirable to minimize inventory levels commensurate with a given level of service.

The basic policy decisions which have to be made in an inventory control system relate to (a) the amount of safety stock which should be kept for an item, (b) when the new production or purchase orders should be placed, and (c) the quantity of production/purchase orders. It is desirable to maintain safety stock so that the possibility of stock outs can be minimized. The quantity of safety stock maintained will vary according to the service level which the company provides. The level of investment in the inventory increases very rapidly as the service level approaches 100%. A compromise between the stock level and the service level has to be made. Statistical forecasting routines such as exponential smoothing can be used to estimate the demand for the item over the item lead time, and the required re-order level can be determined. Alternatively the average usage of the item, over a historical time span, can be calculated; this enables the management to specify the estimated usage of the item over the expected lead time.

The policy to be implemented is always decided during the inventory analysis/planning phase. The type of analysis carried out by most companies is known as the ABC analysis in which the inventory items are separated into different categories. The categorization is based on the monetary value of the items as well as the actual use of individual items. Once the individual items have been grouped together the appropriate safety stock, re-order level and re-order quantities can also be decided, according to the type of control which should be exercised on items included in different categories.

A good computer-based inventory control system must be able to analyse the inventory so that the management can review its policy at more frequent intervals than is possible with manual systems. The system should provide for this analysis to be carried out on the basis of investment in inventory as well as the net return on the investment for the item. Such ABC analysis reports, when used with 'where used' lists (BOMP system) enable the management to purge out some of the rarely used items and make any necessary changes to the company policy.

Following the determination of the safety stock level it is necessary to establish the re-order level, i.e. the level of inventory at which a new order must be placed so that the item does not become out of stock, and the desirable service level for the item can be maintained. The re-order level depends upon the following factors.

1   The lead time for the item.
2   Expected demand for the product over the lead time.

It might be highly desirable to include a safety lead time in the calculations so as to safeguard against the possibility of delayed delivery. An additional feature of an inventory control system is the ability to monitor the lead time whenever a new order is placed; an updated lead time can be determined by smoothing the previous value with the latest one. The use of such a facility might not be realistic with a manual system but the ability of the computer to keep track of large volumes of data makes it possible to introduce this refinement into the inventory control system.

Having established the re-order level it is necessary to determine the re-order quantity. This re-order quantity might be the economic order quantity which attempts to minimize the set-up cost or the cost of placing an order, and the cost of carrying the inventory. The annual use of the item might be based on historic usage or it might be based on the forecasts of future demand for the product. The computer system should include routines for calculating the economic order quantity using patterns of demand which are not based on the stable demand assumption.

Before a proper inventory control policy can be established it is necessary to analyse the data; with manual systems adequate data is often unavailable in a suitable form. Therefore the implementation of a planning phase often has to be postponed until enough data becomes available. The data required for the planning phase can be collected by first implementing the execution phase of the overall inventory control system, also referred to as inventory recording. Inventory recording systems process transactions relating to the issue and receipts of individual stock items. It is frequently necessary to adjust the stock to take account of pilfering or misplacement of some of the stock items. The data input to the system has to be very carefully validated in order to ensure the integrity of the database. A good inventory control system is ultimately dependent on the accuracy of inventory records created as a result of the processing of transactions. If the inventory files are corrupted then the inventory analysis will be wrong resulting in formulation of inappropriate policies. A computer-based inventory system is as prone to the introduction of errors as a manual system unless provision is made for the careful editing of input data.

In addition to the normal input data validation procedures, the following specific checks can be carried out on inventory data.

1   The system can check that the quantity of material issued from the stores is not in excess of the permitted issue level as recorded on the requisition list.

2   Similarly as purchase orders are received the system can check that the items received are in line with the released purchase order, and that the quantities of individual items do not exceed the quantities for which the order was placed. Also if the quantity received is less than the ordered quantity, then an exception report for management attention and action can be produced.

3   A validation program can also check that the data input to the system is reasonable; for example, the transaction should not result in a negative free stock balance.

Inventory transactions data are handled by a large number of people and errors can be introduced at a number of stages during the handling of this data. Since inventory data are recorded in one of the main files, i.e. the part master file of the database, the effect of any errors in the inventory data will quickly spread to a number of other functional areas. It is particularly important that with a computerized inventory control system an attempt should be made to maintain accurate records thereby increasing the confidence which the company employees have in the system.

The effect of inaccurate transactions and records can be particularly severe with a computer-based inventory control system in which the safety stock levels have been reduced to the minimum required for the desired service level.

A computer can be used to keep accurate inventory records relating to thousands of items, provided, of course, that adequate control and supporting procedures are also introduced at the same time. Such supporting procedures will require, for example, that stores are kept locked, not to prevent loss of materials but to ensure that only authorized items are issued from stock. Only in such a system will it be possible to assume that the data files accurately represent the actual situation on the shop-floor or in the stores area. If stock items are not properly controlled in order to ease their availability the inventory records will never reflect the actual situation, which can only be assessed by spending a large amount of time and effort. Similar control procedures are required to ensure that the items received into stock are always placed in the designated bins.

Another function of the inventory recording phase is to prepare the necessary documents which suggest that orders should be placed once the inventory on hand falls below the re-order level for the item. It is highly desirable that these documents are vetted by an experienced purchasing clerk before the orders are actually released.

It is of over-riding importance in an inventory control system to ensure that all the transactions are actually reported, validated and processed by the system. This can be achieved by entering data into the system at the point of origin of the transaction. As remarked earlier v.d.u.'s etc., can be used to input and immediately validate the transactions data. The direct input of transactions data is difficult in the case of batch processed systems. The data may be encoded on tapes, floppy discs, cassettes, discs, or punched on paper tapes, etc. However, the validation of this data, and the reporting and correction of errors encountered will have to wait until the data has been edited. With such a batch system it is important to maintain a file in which a record of errors, encountered during the editing process, is kept. This is necessary so that when correct transactions are entered they can be deleted from the error file, and the attention of management can be drawn to those errors for which correct transactions data have not been entered. In some cases this difficulty of entering and validating data can be overcome using intelligent terminals and the edited data are stored on cassettes or floppy discs, and batch processed at a later stage.

The management could also decide that the material will only be issued against computer generated punched cards, or other documents which can be input to the system using, e.g. optical card readers. However, this policy can result in problems if these documents are lost or if additional material is required. Management will have to lay down the policy after a careful consideration of all the relevant factors. The integrity of data files can only be ensured by going through such detailed procedures. Failure to ensure this datafile integrity will result in the following difficulties.

1   A higher frequency of material/part shortages.
2   Maintenance of high stock levels to ensure that stock outs are kept to a minimum.
3   Need to carry out physical inventory checks at shorter time intervals.
4   Necessity to ensure that items required for assembly are available in store well in advance of the start of assembly operations.

In practice, however, some errors will always creep into the records. It is essential to maintain an audit trail (list) of transactions. Such an audit trail might include the minimum data required to identify the transaction. Typical audit trail data will show the part number to which the transaction relates, the type of transaction, fields affected by transaction, and the person responsible for creating that transaction. This last item of information makes it possible to assign responsibility for errors, and ensure that the likelihood of errors is minimized.

No system, whether manual or computer-based, will always reflect the true situation pertaining to the inventory maintained by the company. Irrespective of how rigorously the disciplines and procedures are enforced, it is unavoidable that some data will be missed out unintentionally or maliciously. The locking of stores can reduce, but not entirely eliminate, the incidence of loss of parts by pilferage. Editing procedures which check the data for all the possible error permutations are extremely difficult to develop. The cost of implementing such an all embracing data validation system will also be prohibitive. It is, therefore, necessary to count the inventory physically so that the inventory levels, as reflected in the database, can be reconciled and adjusted to take account of unprocessed transactions. The computer can be used to help carry· out perpetual inventory counts.

In manufacturing organizations the purchasing department plays an important role since the achievement of the overall manufacturing plan is often dependent upon the availability of purchased raw materials and components which can form a high percentage of the items in the company's inventory. Many production problems have their origin in delays in deliveries and rejection of items received from outside suppliers. It is desirable to ensure that items are purchased at competitive prices from one of a number of possible suppliers. However, the item cost is only one consideration; product quality and delivery performance are other important factors. Thus in the case of rush orders it might be necessary to pay a high price for early delivery. Although many companies have separate purchasing departments, in general the purchasing function can be regarded as part of an overall inventory control system.

Computers are used to generate purchase orders from purchase requisitions. Information relating to purchase orders for individual items can be kept in the purchase order file. By chaining together the preceding and succeeding references to the same part number a full record of the part can be kept and easily retrieved. The purchase order file can be kept up to date by processing information relating to newly received orders, amendments and cancellations to existing orders, receipt and delivery of goods. The purchasing information can also be used to update the performance of individual suppliers. The buyer can retrieve the necessary information from the system and decide about the best source of supply, for a particular item, taking into account the required delivery performance.

Sources of Inventory Transactions

In an integrated system the transactions entering an inventory control system and the part master file can occur in a number of company departments. Let us consider some of the major sources of transactions.

1   The stores department issues raw materials, components etc., so that production work can be carried out on the raw material or the components can be assembled to form the finished products. The issure of items affects a number of fields on the part master file, for example the quantity in stock, the quantity allocated to planned orders, the requirements for various items, and these fields have to be updated. Similarly the receipt of goods into the stores alters fields such as quantity in stock, purchase order quantity, and production order quantity. As the quantity of individual items is physically counted, in response to a computer produced inventory check list the adjustments data are input to the system so that the database represents the actual inventory levels.

2   The sales department personnel, in response to definite customer orders, enter data into the system to allocate free available stock or work-in-process to

particular customers. In the case of items used for the dual purpose of machine building as well as spare parts, it would be desirable to include an additional field in the part master file so that it is possible to keep a record of any provisional allocation of the quantity in stock or being manufactured to customer enquiries. It is thus possible to ensure that the production planning personnel do not allocate the same quantity for a machine building program.

3   When the shipping department withdraws items to dispatch them to the customers, the fields relating to allocated quantity as well as quantity on hand have to be updated.

4   When an order is placed for a given item the purchase department will update the part master file to indicate that an order for a given quantity has been placed with a particular supplier. The date of order is also recorded so that the progress of the purchase order can be monitored. If the order has not been received by the due date then the system can generate a chase list for management attention.

5   Similarly when an order is released to the shop-floor the status of the planned production order is altered. The planned quantity becomes an order quantity. In an integrated system this information can be used to produce requisition lists automatically which are then used to withdraw materials from the stores.

6   The master production schedule, based on definite customer orders or forecasts produced by the sales forecasting system, shows the requirements for finished products or spare parts. These requirements are reflected in appropriate fields on the part master file. The requirements for raw materials, components, sub-assemblies, etc., over the planning time horizon, are generated as part of the requirements planning system to update the relevant fields in the part master file.

7   As items are scrapped on the shop-floor, due to non-compliance with quality control standards, it is necessary to issue additional components and the 'quantity on hand' field of the part master file has to be updated.

8   Following the inspection of items received from outside suppliers the relevant fields are updated. There is usually a delay between the receipt of goods, and the inspection to decide their acceptability or otherwise. In some cases it might be desirable to enter information about the receipt of goods immediately after they have been received so that personnel in other departments, waiting for these items, can, if necessary, expedite their inspection. In such cases it would be necessary to include additional fields in the part master file.

Fig. 13.1 illustrates the typical transactions which take place in an inventory recording/control system.

Computerized Inventory Control System Modules

An integrated computerized inventory control system comprises the following essential program modules:

1   ABC inventory analysis programs.
2   Order policy recommendation programs.
3   Inventory status updating and ordering (Inventory recording) programs.
4   Physical inventory check and validation programs.
5   Purchase order processing programs.

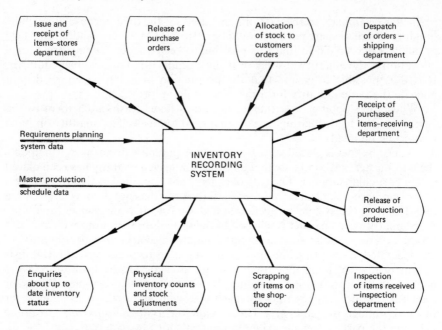

Fig. 13.1 Typical transactions in an inventory recording system

Fig. 13.2 shows the flow of information, and data processing programs used in an integrated inventory control system. Each of these modules is now considered in detail.

ABC Analysis

ABC analysis, based on Pareto's Law, is a management principle used for the development of an effective inventory control system. The inventory items are split into a number of categories. These categories can be based on the value of the inventory and the usage of the item. The value of an item would be the purchasing cost in the case of bought out items. In the case of manufactured items it might be the cost of manufacturing the item. Alternatively the ABC inventory analysis report might be based on the actual up to date product cost, the average cost of the product, the standard cost, or the sale price of the item.

The time period used in calculating the inventory values can also vary. The inventory investment may be based on the demand for the product over a specified fixed period, for example one year, or a variable period, for example from a specified date in the past to the current date.

ABC analysis can also be carried out to take account of the contribution made by the individual items to (a) profit, (b) annual turnover, (c) overhead recovery.

The total investment in inventory can be determined, according to rules specified by the management, by multiplying the item usage and the relevant unit cost. A suitable SORT program can then be used to rank the items according to the level of investment in the inventory. All the inventory items can be divided into the required number of categories. A single parameter value is then assigned to all

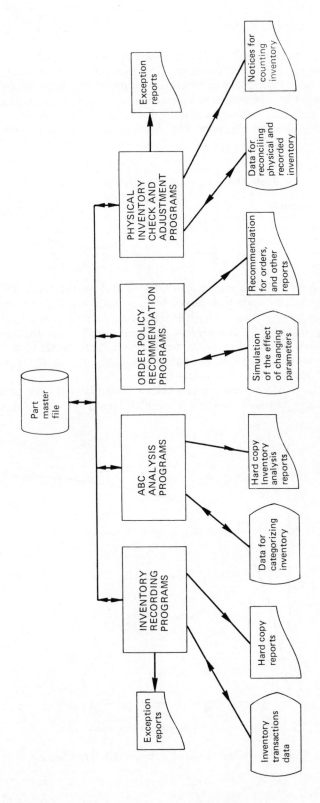

Fig. 13.2 Information flow in a typical inventory control system

the items in a particular category. The total value of an item as well as a complete category of items can be determined. The percentage of inventory value accounted for by an individual item or a category of items can be calculated. Once these values have been computed the management can decide the type and level of control to be exercised over items in individual categories. Table 13.1 shows a typical ABC classification. The same ABC classification is graphically illustrated in Fig. 13.3.

Table 13.1   A typical ABC classification

Category	A	B	C
Items	20%	35%	45%
Money value	75%	15%	10%

The above ABC classification shows that 45% of items represent 10% of the total inventory, and indicates that the products included in category C are either relatively cheap or rarely used. The ABC classification also shows that 20% of all the items, included in category A, account for 75% of the inventory investment. Based on these figures the management might decide to use the re-order point system for items in one category and the requirements planning system for the items in the other category.

ABC analysis is not a static process which is carried out only once. It is desirable to carry out such analysis at regular intervals so that the control

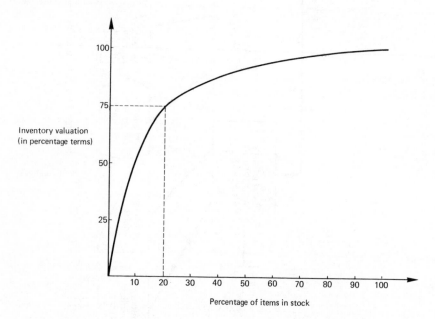

Fig. 13.3 Graphic illustration of a typical ABC classification

exercised over individual items is altered to take into account any changes in the demand pattern.

Order Policy Recommendation Programs

Once items have been split into categories and the management has decided on the type of policy to be used for various categories of items, the next step is to determine the re-order point and the re-order quantity. It is necessary to establish the re-order point in order to prevent the possibility of stockout during the period in which the depleted stock is being replenished. Safety stock is usually maintained to safeguard against unexpected demands and variations (increase) in expected lead time.

It would be ideal to have a situation in which the product lead time and the rate of product demand are constant. But in the business environment it is difficult to assign definite values, with a high degree of confidence, to all these variables. There are far too many factors which affect the situation. Different methods can be used for calculating the safety stock for items in various categories established during the ABC analysis.

The date of order release can also vary. In some cases the management might decide that an order will be released only after the inventory level falls below the re-order point. Alternatively, the date on which the sum of the quantity on hand and quantity already ordered is expected to fall below the re-order point, taking into account the planned or predicted usage of the item, can be calculated. By subtracting the appropriate lead time the order release date can be determined. In the case of dependent items, i.e. the items whose demand depends upon the requirements for assemblies and sub-assemblies at a higher level, the order release date is calculated by subtracting the lead time from the due date.

Safety Stock Calculation

The safety stock can be determined in a number of ways. The simplest method is to calculate the quantity of stock, based on past demand, which might reasonably be used in a specified time period. Thus if the lead time for replenishing the item is A weeks, then the management might decide that the safety stock will be the amount used in AB weeks where B is a fraction of the lead time.

$$\text{Order point} = \text{Demand in one week} \times \text{Lead time } (A \text{ weeks}) +$$
$$A \times B \times \text{Demand in one week}$$
$$\text{and Safety stock} = A \times B \times \text{Weekly demand.}$$

Alternatively, the safety stock can be computed using historical data relating to the forecast and actual usage of the item. The mean absolute deviation (MAD) over the past few periods can be calculated, and the product of the safety factor and mean absolute deviation gives the safety stock. Using this second method it is possible to take account of the level of service which the management wishes to provide for different items. The level of service refers to the percentage number of occasions on which the demand should be met from stock.

The safety factor required to provide a particular level of service can be determined using Table 13.2.

Table 13.2   Relationship between level of service and the required safety factor

Level of service	Safety factor
50%	0
78.81%	1.00
84.13%	1.25
94.52%	2.00
97.72%	2.50
99.18%	3.00
99.87%	3.75

The product of mean absolute deviation and 1.25 is approximately equal to one standard deviation. It is desirable to use Table 13.2, in preference to the calculation of standard deviation, which involves relatively slow computer operations such as addition, division and determination of square roots.

The calculation of safety stock, based on statistics of expected usage, actual usage, and the level of service, can best be illustrated by means of a simple example. Table 13.3 shows the forecast demand, and the actual usage over a number of periods.

Table 13.3   Illustration of the calculation of actual and absolute deviations

Period	Forecast usage	Actual usage	Actual deviation	Absolute deviation
1	210	200	− 10	10
2	210	200	− 10	10
3	220	240	+ 20	20
4	190	200	+ 10	10
5	200	200	0	0
6	230	220	− 10	10

$$\text{Mean absolute deviation} = \frac{\text{Sum of absolute deviations}}{\text{Number of periods}}$$

$$= \frac{10+10+20+10+0+10}{6} = 10$$

Average actual demand $= 210$

Average forecast demand $= 210$

Assuming that the demand forecast for the next period is 210 units then the amount of safety stock which should be kept to satisfy a given level of service can be easily calculated using Table 13.2. Table 13.4 shows the safety stock levels for various levels of service.

Table 13.4 Illustration of calculation of safety stocks for different levels of service

Level of service	Safety factor	MAD	Safety stock
50	0	10	0
78.81	1.00	10	10
84.13	1.25	10	12.5
94.52	2.00	10	20
97.72	2.50	10	25
99.18	3.00	10	30
99.87	3.75	10	37.5

As can be seen from the above example, the management will have to decide the level of service and then invest in the inventory accordingly. Assuming that the management wishes to provide a service level of 99.18%, and a new order is placed immediately after the re-order point has been reached then 30 units will have to be maintained for safety stock. But with a service level of 94.52% and the associated safety factor of 2.00 it is only necessary to keep 20 units of safety stock. This also means that on 5.50% (100 − 94.50) of occasions the demand for the product may not be satisfied before the production/purchase order is received.

The major advantage of using a computer for this work lies in the fact that safety stock can be calculated anew, based on the latest forecasting errors, thereby providing maximum protection, consistent with the desired service level, against stockout possibilities.

The order point can now be computed by calculating the usage of the item, over the replenishment lead time, and adding it to the safety stock.

Calculation of Order Release Dates

Order release dates can be computed by continuously monitoring the transactions as they occur and when the stock level falls below the re-order point level, as recorded in the part master file, the computer can suggest the placement of an order with the usual supplier. The quantity will vary according to the rules specified by the user. Fig. 13.4 graphically illustrates the use of re-order point techniques.

Statistical forecasting routines can also be used for calculating order release dates. The expected usage of the item can be determined, using these statistical routines, and taken into account in the placement of orders, thereby minimizing the possibility of stockout. The main advantage of using this type of inventory control is that by planning orders in advance, time is available for reviewing orders and it is possible to take account of variations in demand. This procedure is illustrated in Fig. 13.5.

When time is available to review a planned order the changes in demand pattern between the planned and release dates can be taken into account and the release date revised if necessary. Similarly if the expected demand will result in a stockout situation as shown in Fig. 13.6, the need for expediting action can be highlighted by an appropriate exception report.

In such a system, used for independent end items, the projected requirements

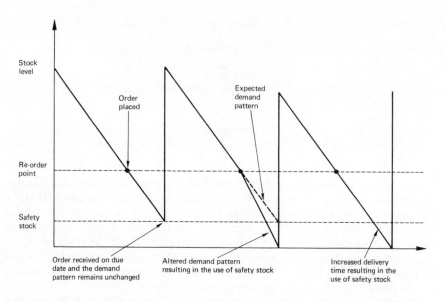

Fig. 13.4 Illustration of the use of re-order point techniques and the need for safety stock

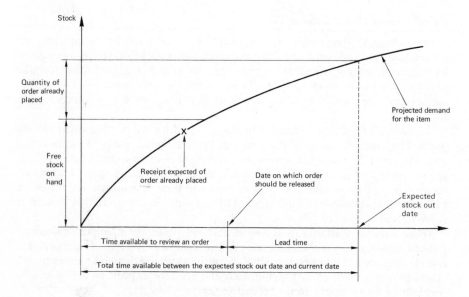

Fig. 13.5 Advance planning of inventory requirements

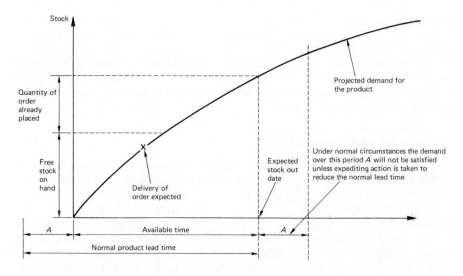

Fig. 13.6 Graphic illustration of the need for expediting action

are first offset against free stock on hand. If the stock on hand is inadequate to cover these requirements the released order quantities are used to meet the demand. If the projected demand cannot be met taking into account free stock and released orders, the order release date is calculated by subtracting the normal lead time from the stockout date.

In general there is some room for expediting special orders in view of the fact that the lead time consists of a number of separate elements as illustrated in Fig. 13.7.

The placement of orders for dependent items whose requirements are determined during the parts explosion–requirements planning process has been discussed in the requirements planning chapter. Very briefly the requirements for raw materials, components and sub-assemblies, over future periods, are determined, followed by a calculation of the date on which the free stock on hand

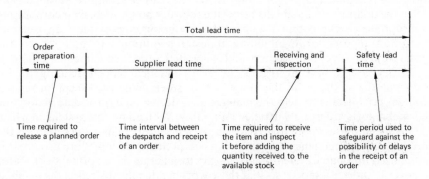

Fig. 13.7 Elements of the total product lead time

as well as the quantity covered by released orders would be used up. The order release date is then computed by backing off the lead time from the expected stockout date.

The actual release of an order may be based on a periodic review of orders, and if the release date of an order falls within the periodic review dates then it can be released for appropriate managment action. In an on-line system the order release may be on a daily basis.

Calculation of Order Quantity

Large orders, placed at infrequent intervals, increase the maximum as well as mean stock levels. Conversely, frequent small orders have an opposite effect. The handling and processing of a large number of orders is expensive.

The order can be based on a fixed quantity specified by the management. Alternatively the order quantity may be calculated using the standard economic order quantity relationship which minimizes total inventory costs. Order quantity can also be specified as the difference between the actual inventory level and a maximum level specified by the management. This method is often referred to as the order-up-to-quantity method. Other possibilities include the use of routines in which the unit cost is minimized taking into account the actual demand pattern as distinct from the average demand assumed in the standard economic order quantity relationship. The frequently used lot sizing techniques were discussed in detail in the requirements planning chapter.

Inventory Recording

Once the basic inventory control policy has been decided by the management the successful implementation of the policy is very heavily dependent on the way in which the record keeping function is performed. This record keeping is concerned with amending the fields in the part master file, so that they reflect the up to date status with regard to the on hand, on order, planned order, and reserved quantities already allocated to customer orders or the machine building program.

During the processing of transactions in an integrated inventory control system, a recommendation for a suggested order can be generated automatically when the quantity on hand falls below the re-order point. Also, by incorporating appropriate checks within the inventory programs, the system can generate exception reports for highlighting unusual conditions. In addition to the generation of exception reports and suggested order recommendations, it is necessary to produce, at regular intervals, detailed reports relating to, for example, the status of all the items held in stock. Whereas exception reports contain brief details to help the management decide on appropriate action, the normal reports are usually detailed and bulky. With an on-line system, in which necessary enquiries about up to date status of inventory items can be made as the need arises, the frequency at which such comprehensive reports are produced is minimized. In a batch system, it is necessary to generate these normal stock status reports at shorter intervals so that these working documents reflect, as much as possible, the up to date status of individual items.

Inventory Recording Routines

In any manufacturing organization the inventory records can be divided into three separate categories.

1 'Planned Order' inventory relates to the quantity for which orders have been planned but not yet released for production or purchase as the case may be. Planned orders are generated as part of the requirements planning system and the quantities of components or raw materials, which must be manufactured or purchased to meet the requirements for end items, are computed. These orders are generally subject to review, to take account of changed requirements, before their release. As the order release date approaches a new order is created following a review of the current requirements. The quantity of 'on order' inventory is increased and the planned order inventory quantity reduced by the same amount.

2 'On order' or 'in process' inventory refers to the quantity (of inventory) which is not currently held in the store room shelves or bins, but is either being manufactured on the shop-floor or is being purchased from an outside supplier. This, however, does not include orders which have been planned but not released.

As soon as the order is released a new production or purchase order record is created. Detailed information, such as the operations completed and the quantity passed on to the next production operation, relating to the orders released to the shop-floor should be kept in appropriate work-in-process files. In addition it is often necessary to keep track of items which are released from the stores for the performance of some operations, and then re-enter the system before being re-issued for incorporation into larger assemblies. In the first case when the item is simply being worked upon it does not lose its identity whereas in the latter case when an assembly is produced the item will not exist independently any longer. One of the objectives of this phase of an overall inventory–production control system is to ensure that the work produced, in the production shops, is in line with the material issued from the stores thus making it possible to maintain strict control over individual inventory items. The following transactions affect the 'in process' or 'on order' inventory.

(i) Receipt of production and purchase orders which result in a reduction in production on order and purchase on order quantities, and a complementary increase in the on hand inventory quantity.

(ii) Cancelled production/purchase orders reduce the production on order and purchase on order quantities.

(iii) The release of new production and purchase orders increases the production on order and purchase on order quantities.

By monitoring relevant transactions the appropriate fields are updated and by building control rules into the system, exception reports for management information and action can be produced if the quantity received is not within acceptable tolerances specified by the management.

3 The 'on hand' or 'in stores' inventory refers to the quantity physically present on the shelf or in bins in the store room. 'On hand' inventory fields are naturally affected by a large number of transactions which take place in the manufacturing environment. The transactions relate to the time span over which items are held after being physically received into the stores before being issued again, whether

to be further worked upon or for shipping to customers in the case of finished goods. The items received into stock may have been worked upon in the production shop and turned into finished goods. In other cases only some of the operations may have been completed on the item, and further work, for example assembly operations, is still to be completed. Alternatively, items may be received into stores following their delivery by outside suppliers and subsequent inspection.

The numerous transactions which affect the on hand inventory fields can be broadly split into two main categories.

(i) Transactions which increase the on hand inventory. These relate to, for example, the planned receipt of production and purchase orders, unplanned receipt of items, return of items, and inventory adjustment transactions which result in an increase in the on hand inventory. Inventory adjustment is necessary when a reconciliation between a physical inventory count and the inventory levels, as recorded in the relevant fields on the part master file, is carried out.

(ii) Transactions which decrease the on hand inventory relate to the planned and unplanned issue of items, and any necessary inventory adjustments.

The task of processing the transactions data may be eased by allotting codes to different types of transactions. The actual processing of data is very simple. When items are received into store the on hand inventory is increased by the quantity received. It would be desirable to include in this transactions processing module the capability to check on the existence of any back orders. If a back order does exist the management is informed that this order, which previously could not be met due to lack of adequate inventory, can now be satisfied.

In the case of stock issue type transactions the on hand inventory is reduced by the issue quantity. The system should include facilities to check whether an issue transaction relates to any existing back orders, and if necessary delete the relevant back order.

The inventory recording module may be used to produce order recommendations, for re-order point controlled items, as soon as the quantity on hand falls below the re-order level. However, the inventory recording module should not initiate orders for other items whose demand is planned using the requirements planning module.

Another type of transaction, which forms an integral part of the inventory control system, relates to the allocation of available stock to particular customer orders or production orders. The end items and spares are allocated to outside customers, while components are allocated to production orders. This reservation is necessary to ensure that items are not used up for other purposes during the period between their planned and actual use. Since the quantity on hand/in process can be allocated to a number of separate orders it will be necessary to keep detailed records of individual reservations in addition to a record of the total quantity reserved for all the orders. Once the quantity allocated to individual orders is actually withdrawn from stock the detailed reservation record is deleted, and the fields containing the on hand quantity and total allocated quantity updated.

If the on hand inventory is inadequate to satisfy a particular order an appropriate back order can be created. When a quantity of the item is received into stock the existence of any back orders can be checked and the back orders satisfied.

The inventory recording module should also include the facility to keep a log of all the transactions processed by the system.

Physical Inventory Count

It is impossible to guarantee absolute accuracy of data input to any system. It is, therefore, necessary, for the success of an inventory control system, to carry out periodically a physical check of the inventory and reconcile differences between the actual and recorded quantities. Following the categorization of items, during ABC analysis, the intervals at which physical inventory checks should be carried out are established. Thus items representing a high level of investment can be counted at more frequent intervals than the items representing a low level of investment. Using a computer-based system the physical checking of inventory items may be carried out on a perpetual basis. Based on the user specified physical count time intervals a computer program can be used to generate automatically, in a random manner, the part numbers of items which should be counted during a particular time period. By specifying minimum and maximum count intervals it is possible to ensure that the same item is not picked too often for counting and also that it is counted at least once during the maximum specified period. The system can regularly check the part master records and determine whether a physical count should be taken. It can also produce the necessary physical count documents for items selected for a physical count. This document might be a form which shows the part number, part description, the on hand quantity as recorded in the master file, and a blank column in which the person responsible for the count can enter the actual quantity. However, to avoid cheating on the part of the counter, it might be undesirable in some cases, to print out the quantity recorded on the part master file. Once the count is completed these documents can be input to the system followed by a reconciliation of these two sets of data. The system program should include a facility to check that the actual physical count, for selected items, is reported back within a reasonable period of time. If the acutal and recorded quantities are not in agreement, or if the physical count is not reported, appropriate exception reports should be produced for action by the management. The system program should be such that each item of stock is physically counted at least once a year.

With manual stock control systems the annual stocktaking ritual is an error prone and laborious process. A major advantage of perpetual inventory check, used in conjunction with a computerized inventory control system, is that it is no longer necessary to shut down the factory, and attempt to count all the items, in a hurry, over a relatively short period of time, which inevitably leads to errors. Also if the same experienced personnel are always used to carry out this inventory check, for all the items, on a rotating basis, the results are likely to be more satisfactory than if people unfamiliar with inventory procedures are used to perform these stocktaking duties. For best results it is also desirable that the physical count should be carried out at the beginning or end of the day so that all the transactions which have occurred during the day are also taken into account.

Purchase Order Processing

The requirements for purchased items are created during the requirements planning phase or as part of the inventory recording module. These requirements in the form of open purchase order requisitions can be converted into released purchase orders after they have been vetted by a purchasing clerk, and a suitable supplier has been selected.

The part master file contains a field which can be used for storing purchase requisitions information. This field can be linked to one of the requisitions for the item. The task of retrieving purchase information can be eased by chaining together all the requisitions for an individual item, i.e. by linking the first requisition for the item to the next. The requisition records would contain information fields relating to, for example, requisition number, item number, quantity required, required date. Other relevant information, such as lead time, cost of the item, on hand stock, planned future purchases, relating to the item is retrieved from the part master file. Within the part master file the address of the first supplier for the item can be stored, and the records of all the competing suppliers for an individual item chained together in the supplier file. Typical supplier record information includes code number, name and address, telephone number, performance rating, and the number of items purchased from him.

A list of purchase requisitions, as determined during the running of the requirements planning and inventory accounting modules, is prepared. This list might take the form of a printed document or it can be displayed on the screen of a visual display unit. The purchasing clerk can then select the preferred supplier by using price quotes and performance ratings data stored on the supplier file. A purchase order can be created and this information stored on the open purchase order file. Typical information in this file would include purchase order number, part number, order quantity, supplier code number, expected delivery date, promised delivery date, unit price, receipt record, inspection record. All the orders for an item can be chained together by linking the first order record to the next one and so on. The order can also be printed for dispatch to the supplier. The performance of the purchasing function can be dramatically improved by using an on-line real-time system in which the buyer can interact with the system and select the preferred supplier by examining their performance rating. Where a printer unit is attached to the user terminal the order may be printed out immediately.

The delivery performance can be improved by following up open purchase orders at frequent intervals. The performance of duties by the purchasing department personnel is considerably eased by including a facility, within the computer system, to monitor the order due date and the quantity due until the order has been successfully completed. For example, the system may be used to prepare a report showing the purchase orders due within a defined time period. Also, in an exception report orientated system a report which shows overdue orders can be prepared. In batch systems a summary report relating to the status of all open purchase orders may be produced to show the progress of the order, i.e. whether an order acknowledgement has been received, whether the supplier has advised about a possible delayed delivery, whether the item is in the loading bay/receiving area, or whether it is being inspected after delivery by the supplier. It would also be desirable to include a facility in the computer system to produce

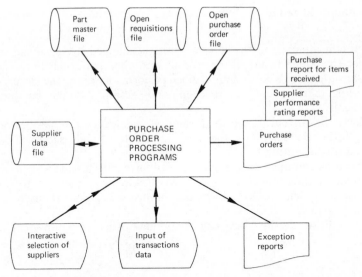

Fig. 13.8 Information flow in a typical purchase order processing system

'expedite/reminder' notices automatically as the order due date approaches. However, before these notices are sent to suppliers they should be vetted, and an attempt must be made to ensure that the data used to produce these reports is up to date.

As purchased items are received in the receiving area, and before they have been inspected, the open purchase order files may be updated. With an on-line real-time system, personnel who are waiting for the arrival of some urgent purchase orders can obtain this information via a visual display unit, and, if necessary, expedite the inspection of these items. With a batch system it would be necessary to prepare reports which list the purchase order items received, and their current status.

The performance data, relating to individual suppliers, can also be updated after the items have been received. This supplier rating should indicate the delivery performance as well as the quality of products received. The delivery performance rating may be in terms of days by which individual items are late. The following relationship can be used to calculate the current rating:

$$\text{Current rating} = 100 - \frac{\sum_{i=1}^{n} \text{DL}_i}{n}$$

$$= 100 - \frac{(\text{DL}_1 + \text{DL}_2 + \quad + \text{DL}_n)}{n}$$

where DL_i = Days by which ith delivery in the current period is late.
n = Number of deliveries.

A higher rating indicates a better supplier performance. Exponential smoothing relationships may be used to determine the overall supplier performance over a long period of time:

New smoothed rating = Alpha factor × Current rating + (1 − Alpha factor)
× Previous smoothed rating

The weight attached to the current rating is changed by specifying suitable values for the Alpha factor. If the current smoothed performance rating for a particular supplier falls below a pre-defined performance rating then an exception report, for management information and action, should be produced automatically.

Similarly the current quality performance, in terms of the number of items rejected during the current period, may be calculated and used to update the smoothed quality rating.

The use of a computer system makes it possible to monitor all the deliveries and keep detailed data about the performance of individual suppliers. It would be difficult to keep such detailed information in the case of manual purchase order processing systems.

Fig. 13.8 illustrates the processing of purchase orders in an on-line real-time system.

Advantages of Computerized Inventory Control Systems

The introduction of effective computerized inventory control systems offers substantial benefits. The major improvements possible by implementing such systems are as follows.

1   It is possible to exercise a high degree of management control over the inventory area provided that the disciplines necessary in a good inventory control system are fully understood and rigorously enforced. Very often considerable benefits accrue due to the use of better discipline and the supporting procedures necessary for implementing a new system. A computerized inventory control system can help in quickly recovering the capital investment in the computer installation. Also, as the user confidence in the system grows, the parallel informal systems, maintained and used by company employees in different functional areas, become unnecessary.

2   It is possible to maintain accurate and up to date inventory records. A comprehensive data input validation program can verify data before the files are updated. This is particularly true in the case of on-line transaction processing systems.

3   The availability of the computer power makes it possible to handle very large volumes of inventory transactions data and any exception reports can be produced immediately to help the managers make decisions.

4   Perpetual physical inventory checks may be carried out without having to shut down the factory (over a short period of time) and count items in a hurry which results in errors.

5   It is possible to value all the inventory maintained by the company on a regular basis.

6   The inventory items can be divided into a number of categories, on a dynamic basis, thereby enabling the management to exercise a high degree of control over expensive items.

7   The clerical effort required to process the large volume of transactions is

reduced. As a result the buyers, for example, can spend more time in efforts to reduce the purchasing costs. Even a small reduction in the cost of purchased items has a large impact on the company profits.

8   It is possible to analyse a large volume of data which cannot be carried out in the absence of a computer system.

9   The problem of dual inventory records maintained by production control and financial departments is eliminated, and the users in different departments use a common consistent pool of data.

10   The problem of inventory obsolescence caused by excessive inventory levels can be overcome or at least reduced. Even with lower inventory levels, and resulting savings, the production delays are reduced by reviewing the stock level at frequent intervals.

11   The inventory can be reduced to the minimum level commensurate with the desired service level.

12   Detailed control of individual items is reduced to a routine computer function and management can devote more time to real problems.

13   In the case of on-line systems the volume of paperwork is substantially reduced. Material handling costs are also reduced due to the fact that personnel in, for example, receiving areas handle less paperwork.

14   By continuous monitoring of transactions the materials/components can be ordered, as soon as the re-order level is reached, taking into account the up to date smoothed lead time for the item and the performance of the individual supplier. The order quantities, re-order points, and safety stock levels can be updated continuously in the light of any changes in demand patterns.

15   It is possible to apply operations research techniques which could not possibly be used in the absece of high speed digital computers.

16   The quality of decision making in all areas is improved as a result of the availability of accurate and timely information.

17   It is possible to react quickly to changes of various types.

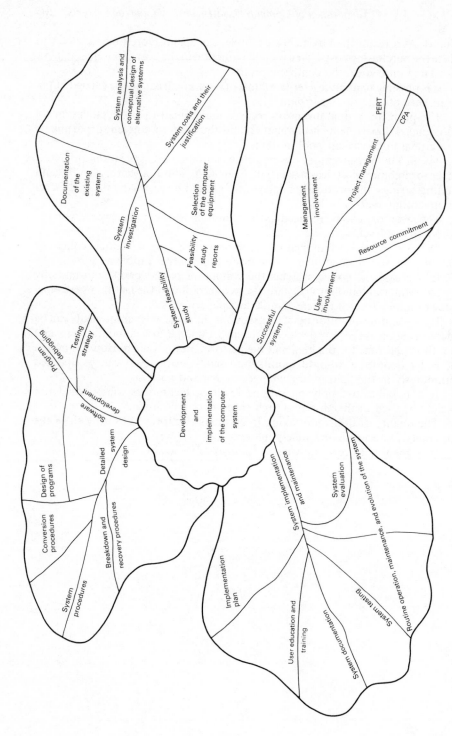

Fig. III.1 Graphic illustration of main subjects discussed in Part III

Part III
Installation of Computer Systems

The successful installation of an effective manufacturing management information system embraces a sequence of activities which includes the identification and detailed definition of the problem, appreciation of the conceptual design of alternative systems, selection of an optimum system, detailed system design and preparation of program specifications, computer programming, testing and 'de-bugging', implementation of the system, routine operation and maintenance of the system. These activities can be completed successfully only if a closely co-ordinated team effort is organized. Personnel from a number of diverse areas such as production control, production engineering, engineering design, sales, purchasing, and data processing have an important role to play in system design and implementation. Above all, production management must play the leading role, and the task of system specification and development must not be left to systems and programming personnel. The project team should be led by a senior production manager who has the necessary authority to make major decisions which might have a profound effect on the procedures used in the organization. If a company employs outside consultants to assist in the system development tasks, the role of the consultants should be strictly defined and the project must always be controlled by company personnel. In this context it is often useful to select some people from within user departments to work with the systems personnel. The departmental personnel become fully conversant with all aspects of the system and are later able to take up responsibility for it's day-to-day running. Once the task of development and implementation has been completed, the system should be fully operational and should require minimum assistance from the computer professionals.

This sequence of activities, illustrated in Fig. III.1, is now discussed in more detail in this section of the book.

14
System Feasibility Study

Introduction

Most companies use an information processing system of one kind or another. The need for a new computer-based system can arise due to a number of factors. The company might have experienced considerable growth resulting in an increase in the number and types of products manufactured. As a result the volume of paperwork circulating within the company increases. Additional people might be required to handle this extra work. A shortage of the adequate information necessary for making planning and control decisions might force the company management to think in terms of possible computerization. Alternatively the management might feel that an existing batch processing system used for production planning and control applications does not effectively serve the needs of the users. Sometimes a production manager may think that the poor performance of a production/inventory control department could be improved by using a computerized system. After witnessing the use of a sucessful computer-based system in another organization, the manager might wish to install a similar system in his own department. More often than not, proposals for computer-based systems come from operations researchers or data processing personnel who have implemented similar systems elsewhere, or feel that computerized production control systems could make a useful contribution towards the task of improving the efficiency of an organization. Whatever the source of the request, the procedure to be used is always the same, and in all instances a thorough feasibility study must be carried out before any attempt is made to install a manufacturing management information system.

The system feasibility study report is used to make all the future decisions. Consequently it is essential to ensure that this study is carried out to a high standard by people familiar with the production planning and control acitivities as well as electronic data processing. Systems analysts or programmers who have previously implemented routine applications such as accountancy or pay-roll systems are not the best people to carry out such studies. A committee consisting of key personnel from affected areas and under the leadership of a senior production manager should be formed to prepare the study report. These people should be seconded to carry out this work on a full time basis. Wherever possible top factory management should also play a leading role in the preparation of the report. Regular meetings should be arranged to review the progress made and difficulties encountered. It might be advisable to ensure that members of the study team keep a log of their acitivities and prepare interim reports so that progress made can be checked against the target completion date.

System Investigation

The feasibility study or system investigation forms the first stage of the system development process. Having considered the possible use of the system the senior management should prepare a brief which summarizes the overall objectives of the system, the time available to carry out this study, and the amount of money which might be available for the computerization process. It is almost always necessary to specify the time available to carry out this feasibility study, otherwise the study will never be completed due to the constant changes in system requirements. The system investigation has to be carried out in order to develop an overall picture of the problems, the possible alternative 'solutions' to the information requirements of the company, and the effort required to implement the selected system solution. The feasibility study shows the possibility of achieving some or all of the system objectives at a given cost, but does not go into details of the data manipulations which must be carried out to perform the required tasks. Also the study should indicate any limitations of the proposed system solution, as well as the interactions between the application under consideration and other sub-systems.

In the past many systems were based on a survey of existing manual methods and the computers simply automated the routine manual procedures. This approach is not always sucessful, and the system designed does not take full advantage of the facilities available on the computer. As a result the opportunity to improve existing systems is lost. An alternative and rather unsuccessful system design approach is to ask users their information requirements. The major drawback of this method is that most users of computer-based production systems, who are either clerks or form the first line of management (that is foreman or departmental supervisors) do not really know their requirements. They only think of the immediate rather than long term problems. Such potential system users are often fearful of the possible adverse effect the introduction of a computer may have on their jobs and are reluctant to participate in the system development process. Other users might ask for the impossible without fully appreciating the constraints within which a system has to be designed and operated.

The best approach is to design a system which can be used to meet the future requirements of the company, i.e. the system should look more to the future and less at the historical experience and current methods of working. The senior and middle management personnel familiar with the future direction of the company should decide about the type of information which they will require to carry out their day-to-day jobs efficiently. The only point which must be borne in mind is that many managers unfamiliar with the operation of computer systems might ask for a large volume of printed information. Experience shows that such systems are very rarely suitable for use in the dynamic production environment.

The system investigator or analyst also requires summary information with regard to the external environment in which the company operates, the administrative, financial, and industrial relations procedures used by the company. This information is necessary because the system solution does not simply relate to the selection of suitable hardware and software; other factors such as current practices, the new computer and clerical procedures, as well as human factors have to be taken into account.

This information can be obtained by discussions with senior management and other key company personnel. As a result the overall objectives also become clear. In some cases it might be necessary to obtain detailed information about the current procedures, so that the cost of the new computer system can be justified. Before authorizing the project and the associated expenditure, the management will wish to ensure that the routine operation of the computer does not result in additional costs without achieving some benefits. In some cases a number of employees might be carrying out simple and routine calculations relating to, for example, the yield of specific processes which can be easily implemented on the computer. The resulting savings can be used to offset the cost of employing additional personnel required to operate the computer system.

Documentation of the Existing System

It is necessary to collect and document all the facts relating to the current operations of the company before performing the system analysis which should reveal possible solutions to the problems. System documentation is necessary because no single person or group of individuals can be expected to remember the details of all the activities encountered during the system investigation process. Formal as well as informal procedures have to be considered. It is necessary to recognize the fact that in most organizations a large number of informal procedures are used in spite of the management's insistence on formal methods. In order to perform their jobs efficiently, some employees often have to obtain access to data which is officially denied to them. All the exceptions to the rule, some of which may be very rare, should also be revealed during the system investigation process. The system documentation should show not just the approved procedures and flow of information but also the actual information flow, the use of information made by various people, and the manipulation of data or information that takes place before being passed on from one person to the next.

Detailed flow charts should be used to depict the flow of information in a graphical manner. These charts illustrate the movement of goods as well as the associated documents and the actions taken in various departments. The work usually involves interviews with personnel from all the affected departments. The investigators should be familar with the techniques of data collection, data recording, flow charting, and system investigation. In addition they should be capable of communicating information and holding interviews with departmental personnel.

It is also highly desirable that company employees be informed of the objectives of the study; where a computer-based system has not previously been used in an organization, the employees are often suspicious about its introduction. In the majority of companies the computerization of the production–inventory systems does not lead to any significant reduction in the number of employees. However, the functions performed by the employees often change. In order to obtain the necessary information and the full co-operation of the personnel, a project must be fully backed by top management and all the employees should be made aware of this backing. The presence of a senior departmental manager in the project team will ensure that the team members get

H

full co-operation from the personnel who operate the current procedures and that any difficulties or misunderstandings are quickly resolved.

A large volume of data is usually collected and used for later analysis. Copies of all the documents used are also obtained and kept in the system study file for later reference. It is often found that the 'facts' collected as a result of interviewing different people are in conflict. In such cases these conflicts must be resolved before proceeding further with the feasibility study. Once all the facts have been recorded in the form of a flow chart, it should be circulated amongst all the affected personnel to confirm that the documented system truly represents the procedures used.

In the case of overall production planning and control systems, the typical information collected at this stage would include:

(a) The capacity and utilization of all the currently available and projected production facilities for the product lines under investigation.
(b) Details of the available manufacturing processes.
(c) Inventory policies, forecasting methods, pricing policies, and histories of various items.
(d) Details of procedures used for order processing, parts explosion, requirements planning, capacity planning, operations scheduling, work-in-process control, and performance analysis systems.
(e) Functions performed by personnel in different company departments.

To summarize, it would be true to say that all the information which might have any impact at all on the system under study should be collected at this stage. It is clear that the study involves the collection of large volumes of data. However, all this information must be obtained before the problem can be fully defined and before an attempt is made to specify an improved system which will function successfully in the particular environment. The information must then be analysed, and the effectiveness with which this step is performed depends on the skills, intelligence, and experience of the systems analysts and production management. The availability of all this information would make it possible to use a 'systems' rather than 'piecemeal' approach and thereby arrive at overall solutions after considering interactions between various sub-systems.

Analysis of Existing System and Conceptual Design of Alternative Systems

Following the investigation and documentation of the current procedures, all the collected data are analysed to arrive at the conceptual design of alternative systems which can be used to meet the overall business objectives. It is important to keep an open mind at the system analysis stage. The objective at this stage of the feasibility study is to identify problem areas and suggest possible methods of improvement. Specifically it is necessary to prepare answers to two basic and related questions, (a) What are the difficulties in the problem areas?, and (b) How can these difficulties be resolved? It is worth noting that in almost all instances there is scope for considerable improvement in the performance of existing production systems, and at this stage it is essential that various methods of improving the current system are fully considered and recorded. Often, the same problem can be resolved by different means. This makes it possible to design a number of different conceptual systems which could meet the requirements for solving a particular problem.

Once the scope for improvements in the problem areas has beeen defined, an assessment of the feasibility of proceeding with the design of improved production systems can be made. Some of the proposed solutions may not envisage the use of a computer-based system. Very often, considerable improvements in the performance of production–inventory functions can be made by employing a better manual system. In any event the introduction of computers has the greatest effect in companies which use good manual systems.

At the system analysis stage all the procedures currently used in the functional area under consideration should be reviewed. Attempts should be made to include all aspects of the current system in the overall concept of the new system. It may not be possible to computerize all these aspects at the same time. However, such overall systems thinking will avoid the frequent and mistaken approach of piecemeal solutions. With a piecemeal approach it might be difficult to interface one system module with other modules, designed at a later stage, without a large conversion and re-programming effort. This opportunity to review all aspects of the system should also be used to make changes to inefficient manual procedures employed in functions which are not to be computerized.

The system analysts must take into account future as well as present requirements. If the future growth and changes are not considered and the system simply automates the current procedure as it is rather than the way it should be, then the system designed will be very restrictive. If the decision is made to implement such a system then substantial modifications to it might be necessary prior to its becoming operational. The system analyst must also consider the long term plans of the company in the field of data processing. A haphazard growth of computer applications and associated procedures will only lead to segregated system elements in which data are duplicated. Such systems also result in a delay in the flow of information.

A factor of overriding importance, during systems analysis, is the mode of operation of the computer system for the application under consideration. Computer generated information must be timely and available when required. As remarked earlier the best results are obtained by using the computer in real-time mode for production or inventory systems applications due to their dynamic nature. However, under some circumstances it might be necessary to use the batch mode even for these applications. In such cases the senior management should be informed about the implications of using the computer in this particular fashion.

It is also necessary to ensure that the proposed system solutions will be acceptable to the users at the operational level. With the currently available hardware and software tools it is possible to implement all the tasks on the computer although some applications are easier than others. The technical feasibility of using the computer for an application only relates to the technique used for implementing it. Many well designed systems fail due to the fact that the abilities of the users to understand the system have been overestimated. Where mathematical models are used in production or inventory control systems, the managers must appreciate the assumptions and limitations inherent in these mathematical models. Only if the system is accepted by the users will it be possible to realize all the potential benefits of the proposed computer system. As an example the system analysts might decide that the operators have to input data to the system, relating to the movement of batches of work and the status of machines, using special purpose shop-floor data collection terminals or via feedback slips which will be read into the system using OCR equipment. The integrity of system files and the ability to produce relevant and timely reports will depend upon the discipline and incentives used to enforce the reporting of this data. In some cases, it might be necessary to link the data reporting with a bonus scheme operated by the company.

The use of the computer to carry out the manual and semi-automatic tasks results in a significant rise in the technological level of the company concerned. The job content of the tasks carried out by company employees also undergoes a change and different skills are required to operate such an automatic system. The company will have to employ people qualified in different skills such as systems analysis and computer programming. The computer system has to be run by skilled computer operators. Once implemented the new computer system has to be maintained and adapted to take account of the changes in the environment in which the system is operating. As a result it is necessary to create new jobs, eliminate others, and change the roles played by some employees. For example, where a computer is used to carry out tasks such as preparation of the shop documentation, the people who previously carried out these tasks will no longer be required. Extra employees will however be required to input transactions data to the computer system. It will also be necessary to assign additional responsibilities to some of the employees. For instance the production manager might have to take over responsibility for efficient running of the new computer-based production system. The system analyst must define the additional tasks which will have to be carried out, and the level of skills required to perform them. The management will obviously wish to ensure that the company employees can be trained to carry out these tasks thereby eliminating the need to bring in people from outside at a higher level which could affect the promotion prospects of existing employees. The latter situation might result in a build up of resentment on the part of these employees who will make every effort to reject and possibly sabotage the new system. Similarly an automation of routine decision making tasks leads to a reduction in the responsibilities carried by middle managers who will not welcome the use of the computer system.

The effect of proposed system solutions on the organizational structure should also be documented. In many instances full advantage of the facilities available with the new computer-based system can only be realized by making changes to the organizational structure.

The gross personnel requirements are determined by analysing the functions to be performed by the system. The manual as well as computer tasks necessary to operate the new system should be listed. This list is also useful for estimating the training requirements.

The times taken to perform individual and total tasks in any envisaged alternative systems, and the possible integration of various sub-systems, represent additional factors which have to be taken into account at the system analysis stage. Due to interaction effects it is essential that sub-systems are not considered and designed in isolation from each other. The possible use of mathematical models to help with production and inventory planning and control tasks should also be reviewed.

The costs of the proposed system solutions have to be estimated. At the systems analysis stage it might be discovered that the management has set its sights too high. Often it is not possible to achieve all of the system objectives within the postulated expenditure. The costs of the system solutions are determined by considering all the needs of the new system. Skilled and experienced analysts can estimate the total system development costs with reasonable accuracy. In this context, it is important to consider the overall system costs and not just the cost of the computer hardware. Hardware is only a small part of the overall system. Software system development costs often exceed the hardware costs. Hidden costs should also be taken into account. These costs relate to factors such as:

(a) System changeover and organizational changes;
(b) Initial setting up of computer files;
(c) Staffing;
(d) Space, heating, air conditioning and lighting requirements;
(e) Costs of normal or special stationery;
(f) Training;
(g) System maintenance contracts.

Where the alternative conceptual system designs have been prepared in sufficient detail, the system operating costs can easily be estimated from the volume of transactions and the mode of operation of the computer.

Cost Justifications for the Proposed System Solutions

The feasibility study report has to include economic justifications of the proposed systems. The requirement for funds to implement a computer-based system will be in competition with those for other factory machinery and equipment, so that similar economic justification must be made. Benefits or cost savings resulting from the introduction of the computer system have to be evaluated. Tangible savings such as reduction in staff levels are offset against the costs of developing and operating the proposed system. Automation of routine clerical procedures such as accountancy and pay-roll calculations leads to a reduction in the number of people employed to carry out these tasks. In the case of production systems there is little, if any, displacement of personnel unless the company has been employing large numbers of people who manually plan and control the production of goods. Therefore the use of computers in such applications may not lead to a substantial reduction of clerical costs. In these functional areas it is necessary to use different cost justification criteria based on factors such as:

(a) Reduction in inventory levels and the resulting savings in interest costs.
(b) Improved customer service levels, thereby ensuring that the company products enjoy an increased share of the market.
(c) Ability to respond quickly to changed circumstances.
(d) Improved and better decision making.

(e) Better control over day-to-day operations.

(f) Ability to expand business without proportional increase in manpower requirements.

(g) Possible reductions in product costs achieved by better planning and control techniques which improve machine and manpower utilizations.

It is possible to allocate economic, or cash values to some of the intangible factors. For example, the reductions in inventory levels realized by implementing a computer-based inventory control system can be estimated and converted into money terms. However it would be difficult to place accurate money values on factors such as improved and better decision making, and the ability to respond quickly to changed circumstances, mainly due to the fact that different people will allocate different values to these factors.

Where skilled company personnel are available, or when outside consultants are used to help with the task of system design, retrospective simulation can often be carried out to judge the improvement which might result from the implementation of a computer-based system. The anticipated effects of a computer-based system are simulated as if the system had been in operation over the previous two or three years, and the theoretical performance is compared with the actual performance over the same period. It is also possible to simulate future events, and any anticipated appreciable changes in conditions over the next few years can be examined so that forecasts of performance under the old system can be compared with the simulated performance using the new system.

It is often difficult to justify computer-based production systems simply on the basis of immediate cost savings. In the case of proposals relating to computer applications in these areas it is often necessary to place emphasis on the long term benefits. Savings might come in the long term due to improvements in the planning and control of the products made possible by the implementation of the computer system. Senior managers who insist on immediate pay-offs and do not think about the long term requirements and improvements in the efficiency and effectiveness of their production systems may jeopardize the future of their companies.

The total effects of introducing the computerized system should be estimated by considering the tangible savings which can be evaluted using conventional discounted cash flow techniques, savings or benefits achieved due to intangible factors, and the possible adverse effects. The last factor is often ignored in practice. It is, however, important to remember that computer-based information and control systems can have adverse as well as beneficial effects on the operations of the department or company. In the case of batch systems, for example, the job satisfaction of managers often decreases.

The senior management has to decide whether to invest money in machinery for producing goods or in a computer system used for planning and controlling operations on these machines. In arriving at its decision the management must take note of the fact that the availability of adequate, timely and reliable information is a pre-requisite for exercising effective control over inventory levels, and for improving machinery and manpower utilization. The total effect of investing the required capital should be considered. Large investments in machinery which is poorly utilized due to lack of adequate information and planning may not be as effective as the investment in a computer-based planning and control system.

Selection of the Computer Equipment

If an in-house computer exists on which spare capacity is available then the applications under consideration will necessarily be implemented on that computer. If a computer has not been used previously or if spare capacity does not exist then it will be necessary to consider the purchase of another system. As the number of application programs executed on a computer increases, the response of the system slows down. In the past there has been a general tendency to use a centralized group computer to carry out all the data processing tasks which have to be performed in an organization. Experience shows that such systems very rarely satisfy the requirements of the users in the production planning and control departments. The very limited service provided and the long turnaround times associated with such systems lead to a frustration on the part of the users. The same criticisms are true with regard to the use of time-sharing bureau services for production control applications. There is now an increasing tendency to distribute computing power where it is required. Mini- and micro-computers are being used to provide such on-line real-time computing service. If the applications under consideration relate to a new self-contained production line, then it is often better to use a new small computer system for carrying out these and associated tasks. The overall criterion which must be used for making a decision about the suitability of any particular computer configuration is whether or not it will provide the service required by the users.

Micro-, mini- and mainframe computers are now being manufactured and marketed by a large number of suppliers. The ranges of the major manufacturers include computers which can perform all the tasks necessary in a production/inventory control system. However, the time-scale of a computer project will often depend on the facilities available on the selected equipment.

At the stage of the conceptual design of alternative systems, investigations should be carried out which enable a short list of potential suppliers to be prepared. Computer hardware, along with the basic operating system, can be purchased directly from the manufacturers. Some manufacturers also supply—often free or for a nominal charge—a number of generalized application program packages which can be used for a variety of purposes, including production planning and control, inventory control, and accountancy applications. However, as remarked earlier, such generalized packages do not always meet the total requirements of individual potential users, and substantial modifications might be necessary to 'tune' the package.

When a broad outline of the proposed system solution has been prepared, the possible computer suppliers are asked to specify a system configuration which may be used to meet the necessary requirements. The customer should give as much information as possible about the proposed system. The following list, although by no means comprehensive, gives details of the factors which simplify the task of specifying a correct computer configuration:

(a) Primary mode of operation of the computer system.
(b) Number of terminals, if any, attached to the computer.
(c) Data volumes envisaged.
(d) File or database organization and type of access to the files/database.
(e) Type, volume and frequency of reports to be generated by the system.

(f) If the use of computer application packages is envisaged, then these applications should be specified.

(g) Secondary devices, i.e. magnetic tapes or discs to be used for storing back-up copies of the data.

(h) Main method of entering data to the system.

(i) An estimate of the computational requirements of the applications to be implemented on the computer.

(j) System growth requirements for the foreseeable future, as well as total envisaged potential of the system.

(k) Required delivery date which in turn is dependent on the target implementation date.

The format of proposals put forward by different manufacturers will vary. However, most proposals will recommend hardware which is suitable for performing the required tasks, the operating system and application packages available for use with the hardware, the peripherals which may be attached to the computer, the cost of the proposed system configuration, the system maintenance support provided and maintenance charges. It is necessary to use a scientific approach in the selection of the best computer system. Irrational choice based on intuition often results in decisions which are later regretted. The senior management and the data processing professionals should prepare a list of factors which should be taken into account in selecting the computer system. It is highly unlikely that any one system proposal will be better than other proposals in all respects. Consequently it is necessary to allocate weights to all the factors included in the system selection criteria list.

Once the manufacturers proposals have been received, they can be analysed to prepare a short list of systems which merit further consideration. Visible record computers and other small computers which only operate in the inefficient uni-programming mode can often be eliminated at this stage. Similarly other computers which cannot meet the specified requirements should not be included in the short list. It is necessary to restrict the number of possible suppliers so that detailed consideration can be given to their proposals.

If qualified technical personnel who can make the necessary selection decisions are not available within the company then help should be sought from computer/management consultants or system houses specializing in this work. Computer manufacturers also offer to help in the task of system selection although their advice is bound to be biased in favour of the equipment manufactured by them. The advice of genuinely independent consultants is likely to be more objective. Some consultants or system houses are sales agents for and committed to the use of equipment manufactured by particular suppliers and are unlikely to be fully objective in the selection of the best equipment.

Considerations in the Selection of Computer Equipment

The price of the basic computer hardware necessary to perform the required functions is only one of a number of important considerations. The low price of a computer system which does not meet the current and envisaged future requirements is irrelevant during the computer selection process. A wrong decision at this stage can lead to disastrous consequences and missed opportunities. The selection of a suitable computer configuration is a difficult and time consuming process. The suppliers never give details of the restrictions which might stop the potential sale of a system. Therefore it is essential that before committing themselves to a particular system the customers should ask appropriate questions and get satisfactory replies. The following factors should be considered at the time of selection of one from a number of possible configurations:

1   Performance of the computer hardware, software and operating system. If a particular computer has been available in the market for any length of time, then other customers or independent consultants will have prepared some system evaluation reports which offer considerable help in the selection of a computer system. Such evaluations should be genuinely independent and not the ones prepared by the computer supplier.

The customer should talk to as many current users as possible. The computer supplier will identify some of them. This procedure is fraught with some dangers. The supplier will always show his best installations. The customer can identify other users by referring to technical computer publications. It is thus possible to talk to these users on their own ground about their experience relating to the use of particular hardware/software, the hardware/software problems experienced during the implementation process, the good and bad points of the proposed system, the level of hardware/software maintenance support provided by the manufacturer. If it is felt that the system performance is unsatisfactory for meeting the proposed requirements, then it should be eliminated from the list of possible contenders.

2   On its own the speed of the computer measured in terms of cycles per second is often meaningless. This speed must be related to the particular applications which have to be implemented on the computer. It is often possible to carry out a benchmark test. A mix of applications, representing the type of problems which will be implemented on the fully operational system, is executed on the computer and the total time required to carry out these tasks on different computers is compared. However, satisfactory and relevant benchmark tests can be performed only if the user can accurately specify a representative sample of the problem mix.

3   Availability of multi-programming, time-sharing, and real-time facilities.

4   The number of user terminals which can be attached to the particular configuration under consideration.

5   The features available on the computer operating system. A good and comprehensive operating system simplifies the task of implementing applications on the computer. The languages available as well as their efficiency, the main memory requirements of the operating system, the capabilities of the utility programs and the ease with which they can be used, the efficiency of the sort routines, and the availability of telecommunications software needed for implementing on-line real-time systems are some of the important software factors which must be considered in the selection of a computer system. A computer system with inadequate software support will compound the difficulties faced by the user organization. An operating system with a standard high level language should be given preference over a system with a parochial language.

6   The file organization techniques which can be used on the particular system under consideration. As remarked earlier some file organization methods are suitable for use with on-line real-time systems while others are not. The company concerned might also be contemplating the use of a database system. In that case the facilities available on the database management system (DBMS) software which can be implemented on the particular computer under consideration will represent a factor of crucial importance. It is always desirable to ensure that the DBMS software is supported by the computer manufacturer. Some of the DBMS software has high main memory space requirements, and when used in conjunction with a very high level query language the overall response time is high.

7   Another important point to be considered is the background of the supplier. This is particularly true in the case of mini- and micro-computers which can be manufactured relatively cheaply. As a result a number of companies without adequate financial strength are operating in this field. A computer system, once implemented, requires continued hardware and software maintenance support from the manufacturer. It is important to ensure that the suppliers will still be in business in a few years time.

8   The level of maintenance support provided by the manufacturer varies according to the type of contract between the two parties and the skills of field service personnel. An adequate local availability of service personnel and spare parts helps to improve the overall responsiveness. The supplier should be asked to give an estimate of the mean time between failures and the mean time to repair.

   With a successful interactive computer system the ability of the users to carry out their day-to-day duties becomes heavily dependent on the availability of the required computer power. A system which frequently breaks down for long periods of time will severely disrupt the normal functioning of the affected departments. Some elements of the computer configuration, particularly the mechanical ones such as line printers, are more prone to breakdown than others. Considerable weight should, therefore, be attached to the overall reliability of the proposed system.

9   The growth potential of the configuration under consideration is an important factor used in the selection of the computer systems. If a particular computer configuration barely meets the initial requirements then it is advisable to use an alternative configuration which can eventually be expanded, because in most cases the requirements tend to increase after a system has been successfully implemented. A configuration which forms a part of a family of compatible computer systems should be preferred to a standard installation. Compatible systems have the same instruction set, but the speed of the computer and the facilities available vary from one configuration to the next. Thus as the number of application programs executed on the system increases and the current computer configuration reaches a saturation point, a new and bigger central processing unit can be used to handle increased work. Thus the upward compatibility of the system under consideration must be checked. It is also important to check that the same operating system or a compatible version of it can be used on the bigger system. As a result the need to convert existing programs and files for use with the new system is avoided.

   If the upward compatibility of the proposed system is not considered at this stage, then the company management might later discover that the current computer configuration cannot adequately meet all the requirements placed on it. In that case the company might have to decide in favour of a new computer with all the costs of conversion, programming, and supporting procedures.

10 If a company has decided to use computer packages to implement production planning and control applications, then the suitability of the packages, which can be executed on the computer under consideration, will represent a major factor used in the selection of the 'best' system. In such cases it will be necessary to examine thoroughly the facilities available on the packages. The management can prepare a list of the desirable features which should be available. The feasibility of using these packages with on-line real-time systems should be investigated. Particular attention should be paid to the amount of effort required to modify the packages. A major modification effort will negate the possible benefits of using a package. Computer application packages often contain errors; the commitment of the manufacturer to provide the necessary software package support should also be taken into account.

11 Another important consideration is the equipment standardization policy adopted by the company management. The use of standard equipment often helps to resolve difficulties encountered at the time of system breakdown or failure. If similar computer equipment exists elsewhere in the company then the urgently required reports can be prepared by running the necessary programs on the other computer. A better service is also provided by keeping spare peripherals which can be used with either of the systems. Where a number of similar systems are used it might be feasible to develop in-house system maintenance facilities.

12 The total cost of the system, including the monthly or annual operating costs, will necessarily play a critical part in the system selection process. The factors considered under this heading should include the expenses involved in the purchase, lease or rental of the hardware and software, system maintenance, employment of people required to operate the system, development of necessary application software, and the requisition of supplies such as stationery. In a batch system the cost of cards on which data are punched and the computer paper used for printing voluminous reports is often high. Another point to be remembered is that in the case of many suppliers the entry point is cheap, i.e. the cost of the basic computer configuration is low, but its enhancement requires increased expenditure. The trade-in price at which the computer might be sold after a specified number of years usage also merits attention.

13 The extent of education and training facilities available for staff at senior management, supervisory, programming, and operating levels should also be taken into account. Consideration should similarly be given to the quality of technical publications and user manuals.

The above list of factors to be considered during the selection of a computer system is by no means exhaustive. In some cases it might be necessary to base the decision on other criteria such as the ability of the equipment to withstand harsh environmental conditions. The decisive factor has to be the suitability of the equipment for performing the required application.

Feasibility Study Reports

Once the existing system has been analysed and the conceptual design of possible system solutions has been considered, these proposals are documented and presented to the management. The format of feasibility study reports will vary from one organization to another. It is important that this study report should be comprehensive and enable the management to decide whether or not to proceed further with the development of the computer-based system and invest a relatively large amount of money. Also if the company has not previously used computers, the contents of this report will be used to decide the possible future commitment to electronic data processing. It is therefore important to ensure that the report does not contain too many technical computer terms and enables the senior management to grasp all the details required to make the decision. A typical feasibility study report should contain some or all of the following features, as required:

1   An introductory section which summarizes the contents of the report.
2   The existing system should be described using narrative as well as flow charts. The flow of information in the current system and its disadvantages should be discussed. The problems faced by managers due to a lack of timely and accurate information should be highlighted.
3   The proposed system solutions for meeting the required objectives are included in this section. For each solution considered the study report should contain an outline of the proposed system, improvements to the existing manual system, advantages/disadvantages of the proposed solution, and an indication of the total cost of the proposed solutions. The resources required for different methods of solving the problem, including the possibility of employing additional people to improve the system manually, should be listed. At this stage only rough estimates can be prepared.
4   The final section should show the recommended system solution and the reasons for its selection. If the use of a computer is envisaged, then the associated computer configuration and its cost, where relevant, should be included in the report. Estimates of the software requirements, total system development cost, additional personnel requirements, time-scale of the system development and implementation process, level of commitment required on the part of people participating in the system development, user involvement, education and training requirements, project responsibilities, and the date by which the system can be expected to go live, should also be covered in this section of the report. A detailed economic justification of the recommended solution, a discussion of the potential for the future growth of the system, and necessary organizational changes are other important factors included in the system feasibility study report.

All the points mentioned above should be adequately covered so that the company managing director or divisional manager responsible for the smooth running of the factory is fully aware of the benefits and problems associated with the proposed solution. The main consideration in the preparation of this system study is that senior managers with a trained mind, who are not experts in the process of system development, are able to follow the logic of the proposals. A clear understanding of the proposals at this stage will reduce the communications problems often encountered during the system development process. The authorization of necessary expenditure will depend upon whether or not the senior management is convinced of the possibility of achieving the potential benefits. The inclusion of all these points in the system study also ensures that none of the important aspects of the system development process have been overlooked, and that the management can be reasonably confident of achieving the targets.

Summary

The major objective of a system investigation study is to determine the technical and economic feasibility of implementing the application under consideration on a digital computer. Once the senior management has specified the overall system objectives, a committee comprising key personnel from affected functions can carry out detailed investigations and determine the alternative means of overcoming the difficulties in problem areas. If the use of a computer is envisaged, then the cost of the hardware/software required to implement the recommended system solution is included in the feasibility study report with other relevant factors such as the time-scale of the system development process, and manpower resources required. The acceptance of the proposals and authorization of the necessary expenditure by the senior management means that the feasibility study report can be used for initiating detailed system design and its subsequent implementation.

15
Detailed System Design and Development of Application Programs.

Introduction

The feasibility study report includes only a broad outline of the proposed system solution. It does not show the detailed procedures which must be implemented to achieve the overall system objectives. Following the approval of the proposal by the senior company management and the affected users, the broad outlines have to be transformed into detailed system specifications. The supporting computer, clerical and fall-back procedures also have to be designed. During the detailed design stage the system designer has to maintain a very close liaison with the user departments so that their requirements can be incorporated into the system. If the system, no matter how sophisticated, does not satisfy the user requirements then it will necessarily be judged to be a failure. A very large number of factors have to be taken into account during the detailed design of the system, which is also used for preparing program specifications. Programs, once they have been coded using the appropriate source language, undergo thorough testing to detect and eliminate errors.

These aspects of the system development are now considered in detail.

Detailed Design

This phase involves a detailed definition of the tasks which must be carried out to achieve the objectives set out in the feasibility study report. The detailed design is critical, because the actual system implemented is based on the work carried out at this stage. Sophisticated programming or the use of complex technical equipment cannot overcome the difficulties created by inadequate system design. The usual procedure is to design a number of small system modules or subsystems which are then linked together to perform the required tasks. The details of the data input to the system, manipulations of the data, and the output reports are also specified. The clerical procedures used to support the efficient operation of the computer system are defined after consultations with the affected users. To summarize, the objective of this phase is to specify explicitly the techniques and tools used to process the data and prepare the information required by the company employees for performing their daily duties efficiently. The detailed system design documents are also used to communicate information to the programmers who prepare the programs which are executed on the computer to carry out the required data processing. It is necessary to consider a number of factors during this phase. The actual factors will vary from one system to another depending upon the application being considered. In general, however, attention should be paid to the following important considerations.

1   The system must meet the requirements of the users otherwise it will not be possible to attain the overall objectives. The designer must avoid the temptation to design and implement a sophisticated system which is barely understood by the users. Therefore he must continuously keep in mind the skills and abilities of the eventual users of the system. In the event of any doubts he should discuss them with the senior functional managers and make sure that he fully understands the difficulties which may arise, as well as the requirements of the users.

2   The philosphy of management by exception should be used in the design of reports generated by the system, i.e. the number of routine reports should be kept to a minimum. Similarly only brief necessary details should be included in such routine reports. More use should be made of exception reports which highlight deviations from expected behaviour. If it is necessary to prepare routine reports containing a large volume of printed information, as for example in the case of batch processed systems, then a separate section of the report should give details of the exception conditions. This will avoid the need to search for such conditions amongst the detailed information, most of which may not be relevant.

The reports prepared by the system must be timely and relevant. Also these reports should not be ambiguous; they should be ready for use, i.e. the user should not have to spend a long time trying to interpret and analyse the information in front of him. He should not have to prepare manually another set of figures from the data made available to him. Wherever possible graphical facilities should be used to depict trends data. The hard copy output should be reduced and more use made of available visual display units etc.

The system design should specify the format of all the reports produced by the system. The potential users of these reports should be identified. Some of these users should be asked to approve the format and contents of these reports. They might be able to suggest some improvements. If possible these suggestions should be incorporated in the system design. The frequency with which these reports will be prepared and the method used to distribute them to the users should also be defined.

3   The routine decisions should be automated so that the company employees can spend more time dealing with exception conditions. However, the temptation to produce a complex and over-automated system by incorporating all the possible exceptions to the rule must also be avoided. Such a complex system would be very expensive. A cheaper and often better solution is to design clerical procedures which can cater for these exceptions. The system should be designed to reject the data relating to these exceptions, and the attention of the user should be directed to them.

4   The electronic data processing system provides information only by transforming the data input to the system. The timeliness and accuracy of this information is dependent on the availability of pertinent data, which must first be acquired. The system design must specify the source and format of data to be input to the system. The techniques used to enter data should also be described. Since data acquisition costs money and ties up useful manpower resources, the volume of data input to the system should be kept to a minimum. This will also reduce the data input errors. Wherever possible the required data should be generated within the system using other data items stored on computer files. This can be achieved by specifying the relationships between different items of data.

It is best to capture the required data at the source of creation. Special purpose data collection equipment or general purpose terminals linked to the computer should be used to carry out this task. This method of working also ensures that data are immediately validated, and in the event of errors the users can be asked to enter correct data.

The following general data validation procedures which check the data for accuracy, consistency and reasonableness should be carried out to ensure that the system files are not corrupted.

(a) The data can be tested for completness. If the input data are incomplete then the system should reject them and request the operator to enter the missing data items.

(b) The logical values of input data can be tested to ensure that they lie within the expected range. Similarly any attempt to enter alphabetic data in purely numeric fields should also be rejected. The number of digits for numeric fields can also be examined.

(c) In many specially designed coding systems a check digit forms part of the code and is used to test the validity of input data.

(d) The relationships between the information stored on system files and the input data items are also used to check their validity.

Although these data validation checks lead to increases in the size of system programs and data processing requirements, they are nevertheless necessary for ensuring the integrity of the database.

It is necessary to assess the total data input to the system. The data input requirements at peak times also have to be estimated so that appropriate support procedures can be defined. For example in an on-line work-in-process control system a large volume of data will have to be input at the end of a shift and unless this is taken into account at the detailed design stage, the system may not be able to cope with the peak workload. The future growth potential of the data input requirements should also be considered at this stage. In some cases it might be possible to use system procedures which ensure that all the users do not have peak data input requirements at the same time.

5   At the detailed system design stage the file organization techniques to be used in conjunction with the system and application programs are also specified. The file organization techniques will vary according to the mode of operation of the applications under consideration. Systems used mainly in on-line real-time mode will have direct access files while mainly batch systems might use indexed sequential files. The file design should be such that new data items can be added, if the need arises, without having to restructure the whole file. The same consistent structure should be used for all the files, and the relationships between items in different files should be defined by means of pointers. The contents of these files, and the lengths of individual fields should also be described in detail.

6   The detailed system design should specify the data processing necessary to produce the required information. Some action is always necessary for preparing reports. In the simplest case it involves retrieval of data stored on secondary storage devices. In other cases it might be necessary to transform this data according to specified relationships. In both cases the files on which the required data are stored have to be defined along with the file organization technique, which in turn decides the type of access to the data. The system design documents used for preparing programs should show the formats of the input to and output

from the system. The requirements for system modules, the interaction between different modules, and the interaction between the overall system and the users should also be specified.

7   The best approach to the design of integrated systems is the use of small system modules linked together by means of well defined and standard interfaces. The self contained system modules should perform specified tasks without any dependence on other modules. A modular systems approach is preferable for several reasons. The total cost of preparing a number of small system modules required to carry out a particular task is often less than the cost of one large complex system capable of performing the same work. Similarly, the total time requirements for developing the modular system are also low. These modules can be tested one at a time before linking them when an overall system test is carried out. The total effort required to test small modules is substantially less than that required for testing a large complex system. The reduction in the complexity of the overall system simplifies its subsequent maintenance and amendment. These modules can also be incorporated in other application systems thereby reducing the total development costs.

8   The system should include facilities not only to link modules within one application, but also for linking a number of application systems. For example there is a natural link between requirements planning and capacity planning systems. Standard interfaces should be developed for connecting different application systems. The production planning and control system, even considered in its entirety, is really a module of the overall business system. The information generated in the production planning and control system is frequently used in other applications, for example the pay-roll module which forms part of the financial system. The need to enter the same data twice can be avoided by using well defined interfaces between different applications. The overall system concept should be carefully thought out. This is not to imply that the whole system will be simultaneously implemented on one computer. In fact any attempt to carry this out will only result in a massive failure. If the overall requirements are carefully thought out and the interfaces between different application systems clearly defined and taken into account at the design stage then the overall system can be implemented in stages over a period of time. The experience gained during successive stages can be used to improve the design of subsequent systems.

9   The levels of skill, quality, potential, and responsibility of the users should also be taken into account during the detailed system design phase. People form an important element of the total system solution. This is particularly true in the case of computer systems which involve considerable man–machine interaction. Such interactive systems are particularly useful in the dynamic manufacturing environment. The users have either to retrieve information from the system or input transactions data. The designer should ensure that the users can perform these tasks efficiently. The need for education and training of the users should not be overlooked.

10   In interactive systems the design of man–computer dialogues is a factor of vital importance. In the current state of technology, computers are unable to recognize the natural languages such as English. Substantial research work is currently being carried out in the field of artificial intelligence which might remedy this situation although it is unlikely to have any significant impact in the

near future. In most commercial applications it will be necessary to program the system so that the user on the terminal gives the expected response. Under certain circumstances it would be desirable to prompt the user about the type of response which might be expected. This, for instance, is true in the case of conversations between the computer system and a senior manager who might very occasionally use the terminal for making enquiries. If the operator has to make a choice from a large number of items displayed on the terminal screen, then this prompting can take the form of asking him to enter the appropriate line number or part number. Such cues are not necessary in the case of an order processing clerk using the system as an integral part of his day-to-day work. In the case of trained dedicated terminal operators whose main function is to input data to the system, the dialogue can be even more sophisticated in an attempt to minimize the data processing requirements. The use of abbreviations and mnemonic codes would be perfectly acceptable in the design of dialogues for such applications. If the dedicated operator is repeatedly entering the same type of data in a fixed order, for example that relating to a number of batches of work released to the shop-floor, then a pre-formatted screen display should be used. The operator need only fill in blanks in appropriate places. This procedure is very similar to the one used for filling in forms. Fields of fixed as well as variable lengths can be used. The maximum expected lengths of variable length fields are specified and the system rejects any attempts to enter longer data items. This procedure is more efficient since it is not necessary to enter data items to the computer one at a time. All the input data are displayed on the screen which enables the operator to perform a visual check and ensure that correct values have been entered in the required fields. Any errors can be corrected before the data are transmitted to the computer. A pre-defined character or terminal key can be used to initiate the transmission of data to the computer. In this context it is important to inform the operator about not only the rejection of the data, but also its acceptance by the system. Any errors should be flagged and an explanation of the type of error should be given. The system design should be such that errors can be easily corrected.

It is not always necessary to carry out a full-fledged conversation between the computer and the user for establishing his requirements. For example if a terminal is solely used for making enquiries about the work-in-process, then the user need only enter the type of enquiry he is making, followed by the pertinent customer name or order number. This would reduce the time required to retrieve and display the relevant information.

The text used in conversation should be clear, unambiguous, and free of difficult and technical words. It is often difficult to fully comprehend a large volume of information. This difficulty can be overcome by ensuring that only a small number of lines are displayed at any one time. Further improvements to the clarity of information displayed are possible by reducing the length of individual lines, i.e. the text displayed should consist of a number of small lines.

The response time of the interactive system is a factor which merits considerable attention. A continuity of the thinking process can be ensured in a system with fast response. Most people do not like to look at an empty screen for long periods of time during which the data are being retrieved and processed. Very short response times are achieved only by increasing the investment in the hardware. In some cases it might also be necessary to use a different technique for organizing the data. Consequently it is necessary to make a compromise between the response time and system cost requirements.

11   The vast majority of computer output consists of fixed format reports. A desirable feature of any computer-based system is the facility to retrieve information stored in the database, in the desired format, according to the requirements specified by the user. The system designer should endeavour to provide this facility so that the user can formulate his problem by interacting with the system. Such a facility goes a long way to reduce the frustrations of the users who frequently feel that they cannot retrieve the information contained in system files.

12   The system should have the capability for expansion and adaptation so that new requirements and changed circumstances can easily be incorporated without the need to redesign the whole system. In some cases this flexibility can be achieved by building some redundancy into the system. This ability to incorporate changes into the system reduces the need to use informal procedures which would be necessary in the case of inflexible systems which cannot be modified without a major effort.

13   Errors always creep into system files in spite of all the data validation procedures which might form part of any system. Hardware faults also lead to a corruption of files, which may not be detected for a long period of time. The system should therefore be capable of performing automatic checks at regular intervals on the records contained in the files. System programs can, for example, be used to examine the contents of fields with fixed data values. Where special coding systems are used, the task of examining the data consistency is often simplified due to the presence of a 'check digit'.

14   It is usual practice for the program coding task to be carried out by people other than those responsible for the detailed design of the system. The system designer therefore has to communicate the design requirements to the programmer. To ensure that the programs successfully carry out the required tasks, the specifications of the programs should be detailed and absolutely clear. Program specifications documented according to the standards prepared by the company, a professional association, or a national/international organization should require very little additional explanation. Ambiguous or unclear specifications will only lead to delays in the development of the necessary programs. The programmer will have to consult the designer every time he comes across incomplete specifications, resulting in a wastage of time and effort of both the parties. The programmers should not be expected to carry out the detailed system design tasks not completed by the designer because this often leads to problems. The programmer who is not aware of the overall system picture will carry out the detailed task as he sees fit which may not be consistent with the rest of the system procedures. This might make it necessary to redesign and test these parts of the system.

Design of System Procedures

The introduction of a computer in a functional area necessitates the specification of procedures for supporting the efficient operation of the new system. These procedures are an integral part of the overall system and should be designed by the system designer after consultations with the users who will operate the system. The agreement of the affected users is absolutely necessary to ensure that these procedures are actually carried out during the routine operation of the system. This will make certain that the procedures designed are realistic and take into account the problems which might he encountered and also ensure that users understand these procedures.

The design of procedures is carried out in parallel with the detailed system design and programming phases. A number of different types of procedures have to be designed and they should cover all the contingencies visualized at this stage. There is no doubt that some possibilities will not have been covered and it will be necessary to improvize when such a condition arises in practice. The objective of the procedure design process is to show the relationships between the users and operators on the one hand, and the computer hardware/software on the other. The actual procedures can have a crucial effect on the performance of the new system, and for a successful system it is generally necessary to pay considerable attention to the human factors. Usually a compromise has to be made between the functions to be performed by the computer hardware/software and the human elements of the overall system. The tasks which can be performed efficiently by the computer at a fast rate should be automated, while other functions which are critical and have to be adapted according to the current circumstances should be carried out by the users. Thus, the system operator can start off the computer in the morning or at the beginning of a shift, but the sequence in which different programs should be executed at that time can be controlled by the computer. The unnecessary introduction of a human element in this work will only result in an increased possibility of errors. Similarly the automation of other duties which can easily be performed by people might necessitate the development of complex system programs leading to an increase in the cost and time-scale of the system development process.

The designer has to assess the relative advantages and disadvantages of using the operators or the hardware/software for performing particular tasks. The point to be remembered is that the overall objective is to produce an effective system which can be easily implemented. However efficient a system, if the users do not understand the facilities available on it and the significance of the output reports, then it will not be used to its full potential. The procedures inform the user how to realize maximum benefits from the system. A poorly designed system will require the use of complex supporting procedures and the resulting inconvenience might deter the potential user.

The procedures documented should show:

1   The rules to be used for conversion from the old system to the new.
2   The functions which must be carried out to support the new system.
3   The changes to the procedures used in conjunction with the old/manual system.
4   How to deal with exception conditions?
5   The fallback procedures to be used in the event of the system breakdown.
6   The tasks to be performed to recover from the breakdown.

Conversion Procedures

These procedures are necessary to smooth the transition from the old system to the new. They will lay down the date on which the changeover will take place and the jobs which various personnel have to perform to accomplish this task successfully. In general this detailed planning for conversion will not take place until all the programs have been prepared.

The most difficult part of conversion procedures is the preparation of computer files from existing manual data. It is not unusual to underestimate the time and effort required to achieve this. There are no major problems if the files are simply to be transferred from an existing computer system, but for a computer system which will replace a manual system all the required data has to be manually collected and converted into a format which can be easily used with the computer system. Typical examples of this required information are product structure data or inventory records relating to the currently active items. This opportunity to prepare new files should also be used to clean up the manual files.

The collection and validation of required information is a mammoth task in itself. It is difficult to estimate accurately the time required to achieve this. There is the additional problem that once collected this data has to be continuously updated until the new system starts up. The manual updating of file records over long periods of time requires substantial effort. Therefore the completion of the file conversion procedure should be timed to coincide with the start of the system testing phase, thereby eliminating the requirement to update the manual as well as the computer files. The problem of inaccurate estimation of time required to carry out this function can be overcome by collecting the static data relating to the records and use it for creating the file. A print out of this static information can be used to ensure the accuracy and completeness of the files, and any necessary amendments can be made. The dynamic data relating to the records are added at a later date near to the start of the system testing phase. When the computer file has actually been set up, it should be subjected to a manual verification so that the initial data actually represents the current situation; otherwise the corrupted files will provide inaccurate information and lead to wrong conclusions.

Routine System Operation Procedures

The successful routine operation of a system is dependent upon the performance of a number of tasks. An operating system manual which can be used as a reference by all the people who might require access to it is an essential requirement. It should show in considerable detail the following aspects of the functions which must be performed.

1   Who should carry out a given task?
2   When must it be carried out?
3   How must it be carried out?

These procedures should cover the use of the computer, the associated peripherals, and the required software. Similarly the methods used for entering data to the system, and for the receipt of system output reports should also be detailed.

The user is mainly interested in the retrieval of information which he requires for making decisions. The system procedures should therefore show the type of information that can be made available to him. In addition the sequence of execution of different application programs, as well as the values of parameters required to produce a particular form of computer output report, should also be specified. As a result the user is aware of the capabilities of the computer system and he can make a decision about the level of reliance that he can place upon it.

The routine system procedures should cover the sequence in which various tasks should be performed. The output from one program frequently forms the input to the next program. As a result it is necessary to ensure that a program is not executed before the preceding program. This should be clearly specified. Similarly some routine reports, for example those relating to the yields of particular manufacturing processes, are prepared at weekly or monthly intervals. The schedule relating to the preparation of these reports should be included in the systems procedures manual. There is also a requirement to dump the database periodically onto discs or tapes for back up purposes. The system designer should enumerate the frequency at which this dumping will take place.

The details of how to obtain access to the computer system and the relevant sections of the database should also be prepared. Unauthorized retrieval and updating of the database must be prevented. User identification numbers, project numbers and passwords can be used to restrict access to the system as well as to sensitive sections of the database. Similar procedures can be designed to ensure that development programs, i.e. programs which have not been fully tested, do not obtain access to the database. The failure to apply these procedures rigorously will lead to a corruption of the database. It is also necessary to assure the physical security of the database by locking up discs and tapes on which back up information is stored.

The exception conditions encountered during the routine operation of a system also have to be catered for. The complexity of an application program may be reduced significantly by devising manual procedures for dealing with exceptions. Some of these exception conditions might be very rare. It is nevertheless necessary to document the response of the users to these conditions.

For the successful operation of a computer system it is usually necessary to use a highly disciplined approach with regard to the manual reporting procedures. If, for example, the tickets used for feedback of data relating to the completion of shop-floor operations are not filled and returned, then the database will not reflect the true work-in-process activity state. The scheduling of work for the following shift or day will be based on wrong data. A frequent repetition of such conditions results in a loss of confidence in the system on the part of the users and they tend to disregard the information produced by the computer. Strict procedures should therefore be laid down following consultations between the management, data processing personnel, and the shop-floor representatives.

Breakdown and Recovery Procedures

From a user point of view the system must be highly reliable. The overall reliability of the system depends upon the reliability of the elements which form a part of it, and can be improved by (a) using peripherals with high mean time between failures, (b) building some redundancy into the system. All systems are bound to break down or fail at one time or the other due to the fact that it is impossible to assure 100% reliability of a system. The frequency of system breakdown will vary from one computer configuration to another. The designer must consider possible breakdowns and devise procedures for dealing with them.

Contingency plans have to be prepared for the system breakdown caused by;

1   Loss of the complete database or sections of it.
2   Loss of or corruption of the operating systems software.
3   Loss or failure of application programs.
4   Breakdown of line printer or user terminals.
5   Failure of read/write head of the disc.
6   Failure of the disc drive, central processing unit, and other components of the system.

Data incorrectly transferred from the secondary storage unit to the main memory leads to unpredictable results. If the database is not corrected frequently then the errors will continue to proliferate.

Increased system reliability can be obtained at higher cost, which will have to be justified. In highly critical systems the management might decide to install a complete standby system which can normally be used for program development work. When a system failure is detected then the processing can automatically be switched over to the standby system. The high cost of such an arrangement does not make it practical in most circumstances. A better approach is to ensure the availability of spare user terminals, a line printer and disc drives.

The probability of the corruption of database can be substantially reduced by keeping duplicate files which are simultaneously updated. Such an arrangement is expensive and offers protection against hardware faults only. Any errors in the program will result in errors in both copies of the database.

The main priority at the time of system breakdown is continuity of the production work. The system is designed to serve the needs of the production planning and control personnel and its failure or breakdown must not be allowed to hold up the work on the shop-floor during the time required to diagnose and rectify the cause of the failure. In batch processed systems the users may not even be aware of the system breakdown provided that they continue to receive their routine reports. It might, for example, be possible to carry out the required data processing on another computer belonging to a company with whom reciprocal arrangements have been made. The breakdown of on-line real-time systems is more critical and it is necessary to design specific back up and recovery procedures which are activated in the event of a breakdown. In the first place the user has to be informed about the system failure. Depending upon the type of failure it might be possible to shut down the system in an orderly manner. A message displayed on the screen would inform the users about impending close down. Under different conditions, however, manual procedures would be required. The system operating manual should give details of the necessary action. Similarly if the execution of a program is abruptly stopped and the terminal appears to be out of action then the procedures should outline the response of the users. Should the failure be reported to their departmental manager or the data processing personnel? The main objective is to give a feedback to the responsible people who can then attempt to recover from the breakdown.

The continuation of near normal working in the affected areas can be achieved by regularly producing print outs of the activity state. While these print outs are not up to date, they do enable the users to carry on with their normal duties. Procedures also have to be spelt out to prevent the loss of transactions data during the period of system breakdown; otherwise the database will not represent the true activity state. A record of these transactions should be kept and input to the system when the normal data processing service is resumed. A prolonged breakdown could result in a large backlog. In such cases it might be necessary to use an intelligent terminal for recording data on floppy discs or other similar devices.

For certain types of failure, for example in the case of the failure of a line printer, it might be possible to provide a degraded form of service. The urgently required reports could be printed out on slow speed hard copy units. In general the users prefer a situation of restricted service to the complete breakdown of the system over a relatively long period of time.

In batch processed systems a number of copies of the database are kept to safeguard against the possibility of failure. A commonly used arrangement is to keep three different versions referred to as the grandfather, father and son copies of the database. The latest transactions are processed against the last son version of database. Successful completion of this work results in a new version of the son copy. The original son and father versions become new father and grandfather. The previous grandfather version is discarded and the tape or disc on which this information was stored is reused.

Procedures must be designed to ensure that the data used for system recovery are not themselves corrupted. The designer must be prepared for the failure of the system during the recovery process. If, for example, the system fails when the latest transactions are being processed against the last son version, then it would be necessary to process two sets of transactions against the father version of the database. Before the last remaining copy, i.e. the grandfather version, is used the management must ensure that there are no hardware or software faults. The above procedure used for batch processed systems is already well established.

In on-line real-time applications in which the transactions data are directly input to the system, a breakdown can have unpredictable effects. The database is updated in place, i.e. modified data values replace the previous ones. The transactions data being processed at the instant of failure may or may not have updated the records. It is necessary to verify that the same data, for example that relating to inventory movements, are not processed twice. A database unprotected against such possibilities will contain errors. Some protection can be provided by using the application program to acknowledge the acceptance of the data by the system after the database has been updated. A sequentially increasing number can also be allocated to the transactions originating from each terminal. When the computer service is restarted the entry of data can be resumed from the last accepted sequential transaction number. Similarly a log of transactions processed should also be kept. This log, containing a record of the significant and minumum amount of input data, is used for creating a new copy of the database corrupted as a result of system failure. Copies of previous versions of the database still have to be kept. They provide a basis for starting recovery from the failure by processing the transactions log against them.

The data are usually dumped onto secondary storage devices which can be removed easily for safe keeping purposes. Ordinarily this copying of information, which can take a comparatively long time for large databases, is carried out after normal working hours. The copying of database during a working day would make it necessary to stop access to the information until the dumping is complete. Two basic techniques can be used for this copying. In a logical dump the records are edited and the database reorganized to improve the utilization of the storage space available on the device used for transferring information. In a physical dump the data are copied in the order in which they existed on the original storage medium.

The frequency at which the database is copied varies in accordance with the actual use of information. Rarely used records may be copied at weekly intervals whereas highly dynamic changes to an existing version of the database may be dumped at hourly intervals. This procedure is referred to as incremental dumping.

Copies of the operating system software and the application programs should also be made and stored in a safe place under lock and key. The access to these copies should be highly restricted, and before using the last copy of any system program another copy should be prepared.

Software Development

The development and implementation of reliable software for systems which involve any degree of complexity or sophistication is very demanding. The software reliability refers to the low probability of failure of a system due to errors. Total reliability of any system connot be achieved.

The software development consists of the following sequence of activities.

1 Specification of the required software.
2 Software design.
3 Coding and debugging of the programs.
4 Validation testing.

The above sequence only defines a framework within which the program development is carried out; there is always an overlap between a given stage and the activities which precede and follow it. The time required to develop software is dependent upon a number of factors including the complexity and uniqueness of the program, the source language used for coding, facilities included in the language, the experience and knowledge possessed by the programmer, the possibility of developing the software in the interactive mode of operation, the availability of software testing tools, and the possible use of utility programs. In the case of small application systems all the activities which form part of the system development process could possibly be carried out by the same analyst/programmer. The main advantage of this arrangement is that there is no confusion between different stages of the project. However, most complex applications involve more work than can be performed by one person. As a result it becomes necessary to use a group of people who work under a project leader. In addition no one person can be expected to have detailed and specialist knowledge of all aspects of the system development process. The use of a group of people means that they can contribute their specialist knowledge, thereby ensuring that the available facilities are utilized in an optimal manner.

The programmer has to familiarize himself with the problems before he can start to design the necessary software. He requires specific and clear answers to a number of questions relating to:

1   The objective of the program, i.e. 'what is it supposed to do?', and the data manipulations which must be carried out.
2   The relationships, if any, between this program module and other modules.
3   The files or database used in conjunction with this program, and the data fields affected by the manipulations carried out in the program.
4   The mode of system operation, i.e. batch or real-time, in which the application program will be executed.
5   The computer configurations to be used for the development and routine running of the application program. This factor will decide the size of the main memory which can be used in the program and its execution time.
6   The peripherals used to input the required data to the system.
7   The contents and format of the input data.
8   The contents and format of the output data.
9   The peripherals to be used for preparing output reports.
10   The frequency at which the application program will be run. It would be desirable to optimize the performance of a frequently used application program.
11   The source language to be used for coding the program.

The code necessary to implement the system module under consideration is prepared by transforming the detailed module design into appropriate programming procedures making use, wherever possible, of the available utility tools. The detailed system design shows the detailed requirements of individual modules. The normal programming strategy is to use the program specifications for preparing a diagram which shows the structure of the input data, and the manipulations which must be carried out to transform it into the required information. Once the program structure has been decided and checked then the details are incorporated into the program. At this stage the programmer can make a decision about the feasibility of using macros or system modules which might have been prepared previously. The possibility of using available standard programs for carrying out the necessary calculations should always be examined, the only proviso being that they should be suitable for the application under consideration. If substantial modifications to the standard package are required, then the programmer might have to spend a long time reading and comprehending the details before he is able to make the necessary modifications.

A program flow chart is then prepared to show the logic of information flow. The number of logical errors which might creep into the program can be reduced by carefully examining this flow chart and performing a dry run to make certain that all the data manipulations are performed in the required manner. This procedure also ensures that the number of program runs necessary to debug the program are reduced to an absolute minimum. The flow chart is then converted into program code using the selected source language. A thorough inspection of the code helps in the correction of syntax errors before the program is input to the computer.

Debugging of the Programs

All programs have to be fully tested and verified to ensure that the code is correct and the requirements specified in the detailed design documents are satisfied. The ideal situation is one in which the code does not have any errors. However, almost all application programs of any reasonable length, when first prepared, contain errors. The production of an error free program is a long and time consuming process. It is not unusual for the programmers to spend a high proportion of their time on the testing and debugging of the coded programs. It is a well known fact that most people without adequate programming experience are very optimistic about the correctness of their programs. They are often of the opinion that the program will work first time, i.e. it will not contain any errors, and that if there are any errors they will be easily detected and corrected. In actual practice the exact reverse is often true. It is frequently impossible to detect and correct all the errors in even a medium sized program. The number of paths taken by the program may be so large that all the possible combinations cannot be tested.

Two types of errors are associated with the software development process.

1   Coding errors.
2   Design errors.

The coding errors take the form of wrongly composed statements which are not in accordance with the procedures specified for the high level source language. Coding errors are also caused by the omission of the required colons, semi-colons, commas, inverted commas and full stops. Wrong transcription of the coded programs also leads to errors.

Design errors relate to the logic of the program and are more difficult to eradicate than coding errors. The main sources of design errors are ambiguity and an inadequate definition of the relationships between different sections of the program. Logical errors also occur if all the possible combinations of conditional statements are not fully analysed for their consequences. It is desirable to eliminate these errors during the design phase instead of carrying them over to the testing phase; otherwise some of these design errors may not be discovered until a particular combination of data is used and wrong results are produced. The user unaware of the program error will utilize these results, unless they are wildly out, to arrive at the wrong conclusions.

Most modern operating systems have facilities which simplify the task of debugging application programs. Many language compilers check the code to ensure that the statements conform to the standards of the particular language. References to instructions and data contained in different sections of the program are checked in addition to the format of the statements. An explanation of the errors encountered is provided to help the programmer correct the code.

Some operating systems include facilities for automatically preparing a flow chart which displays the logical structure of the executable statements included in the code. This computer generated flow chart can be compared to the manually prepared flow chart used for program coding. Any inconsistencies between the two flow charts can then be resolved.

Trace facilities are also used to help detect logical errors. The trace is used to maintain a record of the activities which take place during the execution of a program. When a program is aborted due to the existence of a 'fatal' programming error the trace record is printed out; the programmer can then detect the cause of the error. Many systems also include error trapping facilities which are useful for correcting minor errors. The use of these testing aids leads to increased memory requirements in addition to a significant reduction in the program execution speed. Hence, it is desirable to remove them once the program has been fully tested, and is in routine use.

Testing Strategy

The testing of programs is an iterative process. The first computer run shows some errors which can be corrected before the next test run. At this stage more errors may be detected and corrected. It is important to ensure that the program code used for correcting errors does not itself contain any further errors. This procedure is repeated until no further errors can be detected. It might be necessary to repeat this procedure ten, fifteen or even more times before the program appears to be correct. This does not necessarily mean that the program is actually free of all the errors. Frequently there are undetected errors in a program even though it might have been running successfully for a period of time. Shortage of time also results in inadequately tested system programs. Some of the remaining errors may be revealed during the routine running of the program. An organized test plan should be prepared to minimize the possibility of such happenings. This procedure will also ensure that all sections of the program are fully verified; otherwise the lack of available time and resources might result in a neglect of some sections.

The test plan document should show the type of testing to be carried out. The best approach is to prepare the test list at the detailed system design and flow charting stage. All the decision points and branches in the program should be noted. This is followed by the preparation of input test data which covers the range of values likely to be used in practice, and the expected results. Errors are detected only if the use of test data results in deviations from the expected results. Consequently it is essential to calculate the likely outcome carefully. These calculations, although tedious, are very necessary. An important consideration is that the expected results should not contain any mistakes, otherwise the programmer could spend a long time trying to detect the non-existent errors. Any variations in the format of particular data fields should also be covered. It is desirable to associate a set of test data permanently with each program. This set can be used for original validation and revalidation after any modifications are carried out.

The testing of programs using computers which mainly operate in batch mode is not very productive. The normal program turnaround time with such systems is of the order of one day even though the actual computer execution time may be only a few minutes.

The use of interactive debugging facilities simplifies the task of testing programs. Simple conversational terminals such as visual display units or hard copy units are used to enter the program and run it subsequently. The program turnaround time is reduced to almost nil. Any errors detected can be corrected

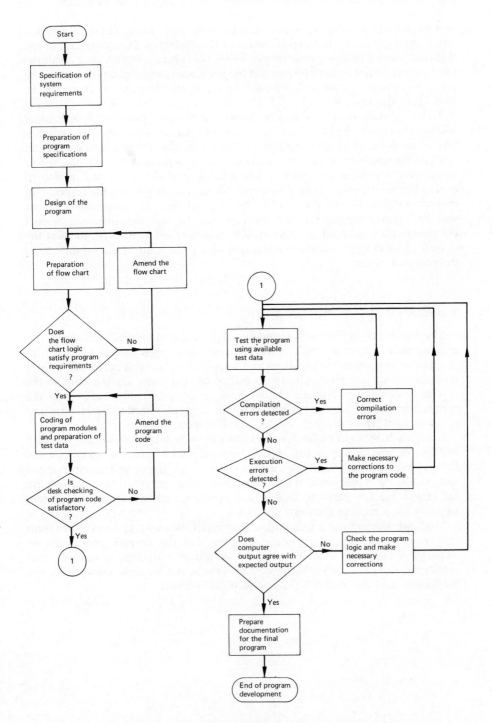

Fig. 15.1 Flow chart of the major stages of the program development process

immediately. In some systems the execution of the program can be interrupted at any stage to examine the current values of the variables. The continuity of the thinking process and programmer productivity is also improved. As a result the total time required to develop and test the program is significantly reduced. The main stages of the software development process are illustrated by means of the flow chart shown in Fig. 15.1.

There are three distinct stages in the testing of a program. The programmer will test the code to make certain that it does not contain any errors. The systems analyst or designer might use different data to see that the program carries out the data manipulations in the required manner. He will check that all the possible conditions have been catered for. The reliability of the software can also be ensured by deliberately using wrong data values and examining the response of the system. Finally the user has to be satisfied that the program meets his specified requirements. He can be expected to test the program against commonly used data values. Abnormally, however, he might wish to 'beat the system'. It is often desirable to encourage this since it might also result in the discovery of errors.

Summary

The detailed design phase is crucial due to the fact that the mechanization of an existing manual system is easy, while the creation of a new system which carries out more than routine automation and improves the overall performance of a manufacturing company calls for considerable ingenuity on the part of the system design team. Human factors have to be taken into account in the specification of work to be performed by the computer hardware/software and the personnel who form part of the overall system.

Adequately thought out and realistic clerical, system breakdown and recovery procedures are as important as the preparation of program specifications. Provided a problem is clearly defined then given enough time and resources it can successfully be implemented on the computer. An experienced designer will use his knowledge and skills to analyse the problem and devise the best system solution for achieving the same end result.

Modular programming techniques, the preparation of a systematic program testing methodology, interactive debugging and the correct application of available debugging tools all result in a significant reduction in the time and money required to develop the system programs. A thorough testing enhances the quality and reliability of the applications software.

16
System Implementation and Maintenance

Introduction

An implementation plan is required to smooth the transition from an existing manual system to a new computerized one. The plan, drawn up after consultations between personnel from all the affected areas, shows the times at which various tasks must be performed for a successful implementation of the system. The details of manpower and their education and training requirements are also prepared. The system as a whole must be thoroughly tested before installation. This is the last stage at which errors can be detected and corrected without a severe disruption of the day-to-day business. It might be necessary to run the old and new systems in parallel for a substantial period until the new system is operating satisfactorily. This procedure implies an increased workload on the personnel from the affected departments and management should be aware of the increased effort required.

The fact that a system has been implemented does not mean that the system development process is complete. Continued effort is required to maintain and evolve the system so that the changed requirements can be incorporated and improvements made in the light of experience gained. The project teams are usually disbanded on completion of a project and people move on to other assignments or companies. In any event no one can be expected to remember all the details of the work carried out during a project. Consequently it is necessary to document the system according to the standards adopted by the company. Let us now examine these facets in more detail.

Implementation Plan

The transition from an existing system to a new one is always a difficult process. All the problems which might be encountered during the implementation of the new system cannot be foreseen. The transition phase starts after the completion of the detailed system design and associated software development, and is completed when the routine operation of the system begins. It is not uncommon to underestimate the time required to accomplish the transition. Accurate estimation is difficult due to the large number of factors involved. This is the stage at which all the elements of the system, human as well as the hardware and software, are brought together in the actual operational environment. This phase is critical due to the fact that all other stages up till now involve concepts or individual components; at this stage the system is complete and has to be operated by the users rather than professional system analysts or programmers. A detailed plan should be developed for replacing the old system by the new one with a minimum dislocation to the routine operation of the departments. The need for a full testing of the computer system as well as supporting procedures must be recognized during the preparation of this plan.

The implementation plan should spell out the sequence of activities which will take place during the transition phase. There is a large degree of overlapping between these activities, for example the detection of errors in a system necessitates their correction before the subsequent activities can be continued. Nevertheless the adoption of a systematic plan reduces the difficulties and confusion which would otherwise ensue. An underestimation of the education and training requirements of the users will only result in future problems. The responsibilities for different aspects of the system change over should be clearly described to avoid the problems of divided responsibility and ignorance. The personnel selected for this work should be fully committed to the use of the computer system and willing to spend time trying to resolve the inevitable problems. Departmental managers or supervisors who resent the introduction of the system in the first place will not be very patient and the teething troubles will be compounded. These people would not wish to take on responsibilities additional to the current ones. The system analysts/designers should also be on hand to sort out any design problems. A system implementation team comprising key personnel who have been trained in all facets of the new system should be formed to reduce the effort required on the part of any one person. The team members can make note of all the difficulties and attempt to resolve them at a later stage.

The management should anticipate the need to shift employees from one position to another. In spite of all their education and training some people, particularly those with hardened ideas, will not accept the introduction of a computer-based system. This is a human problem and must be recognized as such. The affected employees should be moved to similar positions elsewhere within the company, and replaced by people willing to be more flexible.

The implementation plan should not be too rigid. Although it is desirable to meet the target date, the temptation to cut corners to achieve it must be avoided. If the system is not ready or has not been fully tested, then the postponement of the target date is a much wiser decision than implementation of the system on the due date.

User Education and Training

In the past many well conceived and designed computer systems have failed at the implementation stage due to a lack of appreciation by the users of the facilities available and the functions which they must perform. The users were not properly educated and trained regarding the effect of the computer system on the operations in which they were engaged, nor were they informed about the wider aspects of the system. This situation is particularly true in the case of batch processed systems. The users were simply asked to fill in forms or supply data without any explanations of the crucial importance of the accuracy and timeliness of the data required. Similarly large volumes of printed output were supplied to the users, and no information was forthcoming about the alternative formats in which the output could be made available. To summarize the education and training aspects of the system development and implementation have often been ignored. Adequate education and training of all the users and the management is an essential pre-requisite for the successful implementation of a system and as such should form part of the overall system development process.

The actual detailed training requirements of the users will vary according to the scope and complexity of the system and the role played by these users. In addition a general introduction to the overall objectives and broader aspects of the system should be provided to employees at all levels within the company. This acquaintance with the general aspects of the system enables the users to appreciate how the effective operation of the system is dependent upon the completion of specified functions. The senior management will not have the time to attend detailed education and training sessions. They must, however, have an appreciation of not only the broader aspects of the system but also the logic used for transforming the input data into the output information; only then can they ensure a continued satisfactory performance of the system. They must be aware of the decision rules incorporated in the system so that they can be taken into account whenever any changes in operating policy are contemplated. A manager who is not familiar with the details of the computer system will make a decision, the implementation of which could cause severe unforeseen problems. A reasonable degree of knowledge about all aspects of a computer system will ensure that time is made available to adapt it to the changed requirements. If the manager is not certain then he can ask the correct questions before making up his mind.

This general introduction is particularly necessary if there is appreciable difference between the old and new systems. Attempts should be made to familiarize the users with all elements of the system, i.e. the hardware, the application programs, the role of the operating system and the system procedures.

If the system will not be available during particular periods of time, due to preventive maintenance effort, for example, then the users should be fully informed about the reasons for the non-availability.

The contents of the general introduction should vary according to the skills and level of education of the people involved. While the logic of the program can be understood by the supervisory and middle management the shop-floor personnel will in general fail to appreciate it. In their case it would be better to give details of the system input along with a brief description of the transformations required to produce the output information. The users should always be informed about the objectives and significance of the system, the total facilities available on it, and their role in the accomplishment of successful implementation. It is also advisable to prepare a printed hand out which can be referred to at a later stage.

Any additional documents introduced for use with the computer system should be explained together with the reports produced by the system. The comments of the users regarding possible improvements to the system should also be solicited. Every attempt should be made to allay the doubts and fears which the users might have about the system.

This introduction to the system is not a once for all effort. Whenever any major changes are made to incorporate additional requirements or for improving the performance of the system, they should be fully explained to all the users.

Consideration of the training requirements and the actual start of the training program should not be left to a few weeks before the system is expected to go live.

K

The training requirements for some system designs significantly exceed the requirements for other designs. System analysis work at the design stage gives an indication of the training requirements, i.e. the skills in which people must be trained and the scope for shifting people from their current jobs which may not be required when the new system is implemented.

While the necessary procedures can be documented in the form of manuals, it is essential to recognize that most people very rarely read them. The task of educating and training the employees can be simplified by placing emphasis on lectures. Audio-visual aids should be used for imparting instructions to the users. 'On the job' training in the use of terminals etc. should also form a significant element of the total education program. During the development of the system training program it is necessary to take account of the appreciable differences in the capabilities of people to assimilate new information. Therefore this education program should be designed to suit the abilities of different groups of people.

The training of personnel should always be carried out during working hours and should be so timed that there is a minimum dislocation to the routine operations of the department. The people who will have any significant contact with the system, by entering data or using the information produced as a result of data manipulations, should be trained in the use of special or general purpose terminals, the format and significance of the data to be entered and their response in the event of system breakdown or exception conditions. It is always desirable to carry out this training program in co-operation with the training department personnel who can then take over the responsibility for educating new company employees.

Documentation of the System

The failure to recognize the need for adequate documentation of the system through all stages of a computer project frequently leads to severe difficulties. During the evolutionary days of commercial data processing the systems were comparatively simple and the people responsible for developing a particular system carried all the information in their heads or they prepared some notes about the program. Specialization in different aspects of the system development process was comparatively rare and frequently the same people carried out all the tasks. They could make all the necessary modifications to the system. The difficulties appeared only when the person responsible for the project left the company. Such problems and the increasing complexity of systems has created the need for standards to which all the systems can be documented. The adoption of standard procedures means that they can be understood by everyone concerned thereby reducing the difficulties involved in communicating information between different groups of people. If a company does not use any standards then the systems will either not be documented at all, or the documentation will be in accordance with the quirks of the individual designers or programmers. Such arbitrary documentation will fail to convey the necessary information to other readers. Each computer installation should therefore select a set of standards and the management should ensure that they are adhered to. The standards selected should be such that all the information is communicated in an unambiguous manner. A combination of flow charts and narrative is often used to achieve this objective. The use of standards prepared by a national or international organization reduces the need for training people who are new to the company.

System documentation is important at all stages of a computer project. The system design specifications are used as a means of communication between the users and the designers and also between designers and programmers. These documented specifications can also be used at a later stage for the evaluation of the system implemented. A system documented in accordance with specified standards reduces the scope for ambiguous interpretations.

The program specifications, flowchart and the source code should also be fully documented to simplify the task of subsequent modification of the program. An ordinary reader not necessarily familiar with the programming language should be able to follow the program logic by reading the documentation. Some operating systems include automatic flow charters and text readers which can be used to simplify the documentation of the application programs.

The computer and clerical procedures manual should also be prepared in accordance with the adopted standards.

Any changes to the system design, program logic, source code, computer and clerical procedures should be fully reflected in the system documentation manual. The documentation of systems, although expensive, time-consuming and irksome does offer the following major advantages.

1 There is a significant improvement in communication between personnel in different functional areas.
2 Lack of adequate documentation could lead to lengthy delay in making a simple amendment to an existing system. With a fully documented system simple changes can be accomplished within a day or two.
3 The control which the management is able to exercise over the system development process is also improved. The work completed by the analysts and programmers who are forced to document each stage of the project can be easily reviewed. The imposition of strict disciplines also improves the quality of the work carried out.

System Testing

By this stage all the computer programs will have been individually subjected to rigorous testing. However, before implementing a system, all the elements encompassed within it, including the associated manual procedures, should be collectively subjected to a final test. The tests carried out during this phase by eventual system users are based on the assumption that the system is fully operational. These tests should cover all the conditions which an operational system will have to deal with during routine operation. The flow of all the information and documents should also be checked. The system procedures manual should be used to help carry out this task. The fact that eventual users are required to carry out these tests means that they cannot be expected to perform their normal duties efficiently. The management should recognize this and, if necessary, allocate additional resources to complete this phase.

A thorough testing of the complete system at this stage also enables the project team to decide about the need for any additional training, and an estimate of the effectiveness of the system, when operated by user personnel, can be made.

It is inevitable that this testing will result in the discovery of some errors in a system program, or difficulties with the associated procedures. These problems must be removed before the system becomes fully operational. In some cases it might be necessary to alter application programs while in other circumstances the problems could be overcome by changing clerical procedures associated with the system. At this stage changes to procedures are less costly and time-consuming than the changes to programs. In addition to detecting errors and difficulties in the system and associated procedures, the testing by users will also help to pinpoint the areas in which the operating procedures manual does not provide them with adequate assistance in overcoming problems. This is especially true in the case of procedures designed without adequate consultations with the line management.

System errors will be steadily detected all through the system testing phase. The temptation to alter system procedures or programs immediately to deal with the latest error should be avoided. The system must be allowed to settle down for a reasonable length of time and a log of errors discovered should be prepared. This log should then be used as a basis of discussions between the system designers, programmers, line management and user personnel, when a decision can be made about the necessary modifications. These modifications should then be implemented before subjecting the complete system to a new test. In some cases it might be necessary to repeat this procedure a number of times before the system is finally allowed to 'go live'.

System Implementation

The conversion from the old system to the new can commence on completion of the following activities.

(a) Thorough testing of the complete system.
(b) Correction of the errors in computer programs.
(c) Successful conclusion of user education and training.
(d) Detailed documentation of all system procedures.

It is desirable, although not always possible, to start the implementation of the new system at a time of relatively stable or preferably slack activity. This will allow the computer to reach a steady state before there is a need for dealing with large volumes of work.

Subject to the availability of required facilities, it is highly desirable to run the old and new systems in parallel. The period of this parallel running should be just adequate to resolve any minor problems revealed during this period. This will also enable the users to compare the performance of the old and new system. Any employees who are sceptical about the new system can also be convinced and encouraged to gain confidence in the system.

This period of parallel running will also ensure that in the event of any unforeseen problems, the old system can be kept going while attempts are made to rectify them. Lack of adequate resources might make it impractical to use this procedure for large systems. In such cases the system implementation should be phased over a period of time; small sections or modules can be implemented one at a time. Thus the problems in one area can be resolved before the next section of the system is implemented. The sequence in which these small sections are implented will vary from one application to another. In the case of a production planning and control system, for example, the logical sequence would be to introduce the system module for collecting shop-floor data by means of automatic or general purpose data collection terminals, before implementing the next module related to the preparation of work sequence lists. This gradual approach should be preferred to the overnight implementation of a complex system. The large number of inevitable difficulties encountered in the second approach will only serve to convince the sceptical users further about the impracticability of the new system. The users often go by their first impressions and it is highly desirable to minimize the number of problems that they come across. The introduction of a complex system will tend to compound difficulties due to the interactions between different sections of the system.

The senior management, although not deeply involved in system implementation work, can nevertheless provide considerable assistance by being seen to be fully behind the new system. The fact that they are not aware of all the technicalities of the system should not prevent them from providing an encouragement to the line management personnel who will take their cue from the attitude of the senior management. This will also improve the co-operation between the line management and the system analysts; the latter group of people are neither fully familiar with the functions of the department, nor do they have the necessary authority to tell departmental employees how they should perform their duties. The line managers talk the same language as the other users, i.e. there is no problem of communicating the difficulties, which simplifies the task of implementing the new system.

The implementation of a new computer-based production planning and control system often results in substantial changes in the job contents of some of the company employees. Some employees will have to perform functions which are far removed from their previous duties. People expected to perform more demanding tasks under the new system will have to be compensated in the form of higher salaries. New job descriptions should be prepared and subjected to evaluation using agreed procedures. If the system operates successfully but some of the employees find it difficult to adapt themselves to the needs of the system, then they should be shifted to positions elsewhere within the department or company. If a large number of people were employed to carry out clerical functions with the old manual system then some of them might become surplus to the current requirements. A soft approach of redeploying these employees to other positions and subsequent natural wastage over a period of time should be used. If personnel are declared redundant at a time of the introduction of the new system, then the probability of a successful implementation is significantly reduced. Senior management must make these policy decisions before the commencement of the routine system operation.

Routine Operation, Maintenance and Evolution of the System

After the system has attained full operational capabilities upon completion of the system testing and implementation phases, it must be efficiently maintained if the users are to have continued confidence in the information produced by the computer. Strict disciplines have to be enforced to ensure that all the data are input to the new system, and that the old manual system is not used as an informal alternative to the new one. If employees are able to maintain the old manual system, then it would be reasonable to conclude that the work content of their function does not fully occupy their time. The existence of such parallel systems should be discouraged so that the full potential of the new system can be realized. The success of any system can only be measured in terms of the attainment of objectives set out at the feasibility study stage which in turn depends upon the acceptance of the system by the users.

A successful computer system often becomes the centre of all the activities in a department and any breakdown of the computer or the associated procedures can result in a dislocation of these activities. To avoid this possiblity the responsibility for maintaining the system should be clearly spelled out. Similarly the senior management should assign responsibilities for updating the system. Failure to specify these responsibilities could result in inadequate updating and maintenance of the system and all the subsequent problems inherent in such a situation.

The need to inform new company employees about all aspects of the system should not be overlooked. While employees at the direct operator and user level can learn from colleagues or during on the job training, the education requirements of new managers are often ignored. Facilities must be made available to inform them about the objectives of the system and the logic of the programs. They will be able to effect any necessary changes only if they understand how the system works.

The computer manager or operator should keep a complete record of the system operation so that its performance can be assessed. Details of hardware and software failures should be fully recorded. The software errors may be corrected internally if the departmental personnel are familiar with the application programs; otherwise it will be necessary to request the systems analysts/programmers to make the necessary amendments. These amendments should be documented and incorporated in the system documentation and user manuals.

Whenever any changes to the organizational and manufacturing procedures are envisaged, the system designers or qualified departmental personnel should be informed at the earliest possible opportunity so that any necessary amendments to the programs can be made in good time.

Any computer system operates according to the decision rules incorporated in the application programs. In the dynamic production environment, the conditions which a particular system is supposed to reflect undergo continuous changes. If the system is to serve the objectives, then it must be adapted to reflect current conditions. System evolution also takes place as users learn from experience and make suggestions for possible improvements to the programs, the format of output reports and the messages displayed on the v.d.u. screen, or associated clerical procedures.

Other modules for performing additional desirable tasks can also be added following the implementation of the basic system. Thus, the development of computer systems is an iterative process. The management would be wrong if it believes that the system development process finishes after the successful conclusion of the implementation phase. Over a long period of time the latest system design might bear very little resemblance to the first one. The need for frequent or expensive modifications to, and haphazard growth of the system can be minimized by conceiving the overall requirements at the basic design stage. The system should be designed so that new application programs can be prepared and added without the need for complete reorganization of the database and existing system programs.

The original design of a computer system being introduced in a functional area for the first time has necessarily to be simple. Most such systems are based on the concept of 'feedback control' systems, i.e. the computer system provides information only after something has gone wrong so that the management may take the necessary corrective action. When such a system has been operating successfully in an area for a reasonable length of time the management might decide to create a new system based on the alternative and often better concept of 'feed-forward control systems'. In such systems the current data are used for making estimates of future events. It is not necessary to employ any sophisticated mathematical techniques. Simple decision rules incorporated within the program are used to give an indication of the future happenings based on the premise that the current conditions will not change. For example a work-in-process control system can be future oriented if it includes facilities for giving warnings about the unlikelihood of the completion of operations on a particular batch of work by a specified time. The management can then take the required control decisions based on these warnings.

System Evaluation

The success or failure of a system depends upon the initial objectives and the criteria used to evaluate the performance. The original objective might be a reduction in the number of personnel who carry out routine clerical tasks. Alternatively the objective could be the provision of timely information required for making better control decisions. Very often the initial system objectives are overambitious in relation to the resources made available for the system development process. The evaluation yardstick must be established after taking into account the constraints within which the total system was developed.

A follow-up investigation of the system should be carried out after it has been in stable operation for a reasonable period, for example three to six months. Preferably, this investigation should be carried out by personnel other than those responsible for the original system design and implementation. The investigation would reveal whether or not the system is performing as well as was originally envisaged, and if the projected cost savings have actually been achieved. In the event of the performance not being up to expectations, then the causes of the relative failure must be fully investigated and documented so that any lessons to be learned are applied in future system development work. The evaluation report should show the system operating costs, intangible and tangible savings achieved, the accuracy of information processed, and the acceptance or rejection of the system by all the affected personnel. The need to extend the system to cover additional functions should be highlighted.

Summary

The system implementation phase is important due to the fact that at this stage all the concepts and elements of the systems are combined together in the operational environment. A thorough testing makes it possible to correct the inevitable errors in the system before it goes live. Parallel running of the old and new systems, for a period of time, enables the users to compare their performance and slowly develop confidence in the computer-based system. However, once the new system has become fully operational, every effort should be made to restrict the tendency to use the old system as an informal alternative.

Adequate user education and training is essential for the success of any system. A properly structured program should be developed to meet the broader educational requirements of all the employees and the training needs of people who will interact with the system. The training costs will be more than recouped in the form of reduced efforts required to implement the system and its subsequent acceptance by the users.

All systems must be documented to standards adopted by the company. Failure to do this will have repercussions at a later stage when the system must be modified to incorporate changed requirements.

17
The Successful System

Introduction

A successful computer-based production system is one that enables the management to exercise sufficient control over manufacturing functions carried out in the company by making timely decisions; a system in which the activities can be rescheduled, at the earliest opportunity, to incorporate the changes in operating conditions.

The development of such successful computer systems calls for a sustained effort, over a long period of time, on the part of the production management and computer specialists. Only by bringing together these people is it possible to bridge the gap between the two disciplines. The effectiveness of the computer-based production system will depend to a large extent on their ability to understand the full potential of the computer in this area, in addition to an appreciation of the pitfalls which must be avoided.

Management Involvement

Computer applications in the field of production systems have met with a reasonable degree of success. Some of the blame for the relative lack of success lies with user management personnel who have failed to get involved in the system development process. Many systems have been developed by systems analysts who fail to appreciate the problems of the users. These computer professionals might be experts in programming, but system development is far more than mere programming. Software development is only a technical detail which can be carried out by the specialists after the system requirements have been specified. The crucial part is the design of the system and user management must be fully involved during this phase. Left to himself the computer professional will inevitably design and implement a system based on his restricted and often superficial knowledge. Such a system, although efficient in computer terms is ineffective in user terms.

One of the major causes of the lack of management involvement is a fear of displaying ignorance. Another factor is that they are busy with their day-to-day activities and do not have the time to think about the development of new systems incorporating improved operating techniques. Even if they do get involved, the lack of appreciation of the systems approach and the interactions between different systems stops them from making a substantial contribution. Often the systems are developed in isolation and a haphazard growth takes place. The data are also duplicated in a number of application systems. To avoid these possibilities the management at all levels must acquire some knowledge of system technology. It might be necessary to send management personnel on courses where they can learn about computer applications and the system development process. This training away from their usual place of work will enable them to shed their inhibitions about computers.

If the management of a company has decided to implement a computer-based system then it must be prepared to commit its own time, in addition to the allocation of required money and manpower resources to the system development process. A senior manager familiar with the overall business objectives as well as the functions of the departments in which the computer system is to be implemented should be selected to lead the project team. This task must not be delegated to systems analysts. Other managers should also be available to give any necessary advice. An active involvement of the user management will enable them to make significant long term improvements to operating procedures.

A high level of user management involvement is also essential to ensure a smooth transition from an existing system to a new one. Managers involved in the design of a system will want to make sure that it is successful. They will have to operate and maintain the system during its routine operation, and can point out at the earliest opportunity all the difficulties which the system will have to face. As a result the probabilities of design errors, due to wrong assumptions made by the systems analyst, are minimized. The line management will also ensure that the system is user oriented and not computer oriented. A lack of management involvement often leads to a situation in which user problems are ignored in order to achieve higher computer throughput. The efficiency of computer operations, although clearly important, must be a secondary consideration.

Also, the resistance to the implementation of such a system on the part of clerical employees who take their orders from line managers and not systems analysts, will be minimized. These employees often feel threatened by the introduction of a computer system. Some of them, particularly the older ones, are set in their ways and do not wish to see any changes made to existing procedures. Others are of the opinion that the computer will take over their jobs and that they will be made redundant. The co-operation of these employees is vital at the system investigation stage. To achieve this co-operation, the management should perhaps make a clear statement to the effect that none of the existing employees will be subjected to redundancy. This will reduce, but not entirely eliminate, the resistance to a new system. The clerical employees are not the only ones who have reason to resent a new computer-based system. Some of the supervisory personnel only make routine decisions which can easily be automated thereby eliminating the need for their positions. They will make every possible effort to hinder the introduction of a new computer-based system. The presence of a senior company manager will ensure that the investigation team is given the necessary co-operation, although perhaps unwillingly, by all the employees concerned.

The senior managers familiar with the future direction and level of the business activities will also ensure that any subsequent growth in requirements is catered for. Similarly they will endeavour to restrict the scope of the system to activities which will show real benefits. The computer professionals often get carried away and try to produce an all embracing system which will never be implemented in practice.

The line management involved in the system development process should also take a leading role in the specification of manual procedures necessary to support the operation of the computer system. Only they have the authority to alter existing procedures and ensure that the new procedures are realistic and do not infringe any existing agreements with the Trade Unions.

Project Management

In many respects the management of a computer project does not differ significantly from the control of other projects undertaken within a company. Most of the usual project control tools such as critical path analysis, and program evaluation and review techniques can be used to achieve deadlines, provided of course that adequate manpower resources are made available and experienced personnel estimate the times required to complete different phases of the project. Many computer projects are behind schedule because most systems analysts tend to be rather optimistic.

The main difference between a major computer project and other projects is that the computer soon becomes the hub of all the activities in the company. The computerization of the work of one department has an impact on the work of other departments also. For example a computer-based work-in-process control system helps to improve the delivery performance thereby having an effect on the functioning of the sales, marketing and shipping departments. Similarly the work of the production planning and control department is also affected. It is due to these wider implications that the management must exercise a higher degree of control over the progress of a computer project. The capital invested in such a project may not be higher than the expenditure on for example a specialized machine tool, but the introduction of such a machine tool affects only a very small number of employees.

Having decided to use computers in the operations of the company the top management must be seen to be committed. The management can exhibit this commitment by allocating adequate material and manpower resources. In addition to the authorization of necessary capital expenditure, the management can show real interest by selecting competent company personnel to carry out this work. The assignment of personnel must be based on their suitability and no other criteria. The choice of a project controller of considerable merit is of vital importance. He must possess the necessary authority to make major decisions and should have the full confidence of the top management. Too often in the past, managers moved sideways have been put in charge of such computer projects, irrespective of their suitability for the work. Another point to be remembered is that the project controller must be a business manager familiar with the system development process and not a data processing specialist who is likely to take a very narrow viewpoint. The project controller will have the responsibility for coordinating the work of the project team which should include representatives of all the affected user departments as well as systems and programming specialists. Representatives of other departments whose functions will not be computerized in the short term should also be invited at the overall system design stage so that interfaces between different modules, some of which may not be implemented for a long time, are considered and developed.

The top management should show interest by reviewing the progress of the project at regular intervals. If top managers are seen to be taking enough interest, then middle managers and other employees are also forced to learn about the system and get involved in the development process.

It is highly desirable that the whole process of system development from the feasibility study stage to actual implementation is carried out by the same members of the project team. This will result in a continuity of effort, elimination of communication problems, higher quality of work, and minimization of time delays. The only time at which additional people are involved is during the development of software. If the program specifications are documented in accordance with some agreed standards, then there should be no major problems due to this involvement of programmers.

As has been seen in earlier chapters the development of a computer-based system consists of a number of overlapping activities. It is difficult to estimate accurately the time required to complete any individual activity. Past experience and knowledge of the difficulties and problems encountered during the system development helps in the estimation of required time. These estimates are only rough indications and have to be revised in the light of the progress of the project.

The critical path analysis technique may be used to determine the critical path by representing the major activities in the form of a network and using the rough estimates of times required to complete them. The uncertainties in activity duration times can be explicitly accounted for by using the alternative program evaluation and review technique (PERT). The application of these techniques will enable management to decide about the activities to which additional resources must be allocated in order to meet the deadline.

Conclusion

The major objective in the preparation of this book has been to explain, in broad terms, the process of developing computer-based production planning and control systems in manufacturing industries. All the major sub-systems have been discussed in considerable detail. In the wide ranging industrial environment, the relative usefulness of these sub-systems will vary according to the nature of the product, the manufacturing technology required and the size of the company. Similarly the sequence in which these sub-systems are implemented will differ from one company to another in accordance with the problems and difficulties faced by them. Some companies will find it profitable to develop the inventory control system first thereby reducing their operating costs, while others might wish to exercise increased control over the work-in-process before considering the implementation of other systems. However, before getting involved in the development of any individual sub-system, the long term role of the computer in the operations of the company must be defined. This will minimize the possiblity of haphazard growth of computer applications designed to overcome immediate problems in an area. Only top management can decide this role and specify the type of information processing system which must be developed for making improvements to the control of manufacturing operations and the quality of decisions made. User management personnel must be involved in the preparation of detailed system specifications. In this task they will require assistance from computer professionals who are not familiar with the details of manufacturing operations.

It is hoped that the contents of this book will help management in the specification of the type of system required and at the same time enable computer professionals to become familiar with the complexity of the production function, and then work together as a team. The resulting partnership will contribute to the success of the systems developed in any company.

Bibliography

This bibliography, which includes only a small representative sample of relevant publications in this field, is divided into the following four sections.

1   General background reading.
2   Hardware and software.
3   Computer applications and techniques in production planning and control systems.
4   Computer systems implementation methodology.

1 General background reading

1.1   Ackoff, R. L., 'Towards a system of systems concepts', *Management Science*, **17**, 11, 661, 1971.
1.2   Allen, D. O., 'Planning the material and production control system', Proc. 19th annual conference of the American production and inventory control society, Atlanta, Georgia, USA, 214, 1976.
1.3   Appleton, D. S., 'A strategy for manufacturing automation', *Datamation*, **23**, 10, 64, 1977.
1.4   Beishon, J., and Peters, G. *Systems behaviour*, Open University Press, 1972.
1.5   Chestnut, H., *Systems engineering tools*, John Wiley and Sons, 1965.
1.6   Chestnut, H., *Systems engineering methods*, John Wiley and Sons, 1967.
1.7   Dearden, J., Myth of real-time management information', *Harvard Business Review*, **44**, 3, 123, 1966.
1.8   Dearden, J., 'MIS is a mirage', *Harvard Business Review*, **50**, 1, 90, 1972.
1.9   Emery, F. E., *Systems thinking: selected readings*, Penguin, 1969.
1.10 Flagle, C. D., Huggins, W. H., and Roy, R. H., *Operations research and systems engineering*, Johns Hopkins Press, 1960.
1.11 Fogel, L. G., *Human information processing*, Prentice-Hall, 1967.
1.12 Forrester, J. W., *Industrial dynamics*, MIT Press, 1961.
1.13 Grindley, K., and Humble, J., *The effective computer: A management by objectives approach*, McGraw-Hill, 1973.
1.14 Hall, P. E., Holden, G. K., and Green, A. H., *Computerguide 9: Production Control*, NCC Publications, 1973.
1.15 Higgins, J. C., *Information systems for planning and control: concepts and cases*, Edward Arnold, 1976.
1.16 Jenkins, G. M. and Yule, P. V., *Systems engineering*, C. A. Watts and Co., 1971.
1.17 Kochhar, A. K., 'What's happening in the computer revolution', *Director*, **29**, 10, 61, 1977.
1.18 Kochhar, A. K., 'Use of computers in manufacturing systems 7: Trends in

computer systems—The way ahead', *Machinery and production engineering*, **130**, 3344, 50, 1977.

1.19 Orlicky, J., *The successful computer system*, McGraw-Hill 1969.

1.20 Parnaby, J., and Billington, D., 'Computer modelling and control of manufacturing systems with particular reference to tailoring industry', *Proceedings of the IEE*, **123**, 8, 835, 1976.

1.21 Pegels, C. C., *Systems analysis for production operations*, Gordon and Breach, 1976.

1.22 Rickert, J. P., 'On-line support for manufacturing', *Datamation*, **21**, 7,46, 1975.

1.23 Schmidt, J. W., and Taylor, R. E., *Simulation and analysis of industrial systems*, Richard D. Irwin, 1970.

1.24 Stewart, R., *How computers affect management*, Macmillan, 1971.

1.25 Tarrant, J. J., *Data communications and business strategy*, Auerbach Publishers, 1972.

2 Hardware and software

2.1 Barron, D. W., *Computer operating systems*, Chapman and Hall, 1971.

2.2 Bartee, T. C., *Digital computer fundamentals*, McGraw-Hill, 1972.

2.3 Bohl, M., *Flowcharting techniques*, Science Research Associates, 1971.

2.4 Bunt, R. B., 'Scheduling techniques for operating systems', *Computer*, **9**, 10, 10, 1976.

2.5 Chapin, N., 'Flowcharting with the ANSI standard: A tutorial', *ACM Computing Surveys*, **2**, 2, 119, 1970.

2.6 Cronin, F. R., 'Modems and multiplexers—What they do for data communications', *Data Processing*, **12**, 11, 31, 1970.

2.7 Curtice, R. M., 'Integrity in database systems', *Datamation*, **23**, 5, 64, 1977.

2.8 Date, C. J., *An introduction to database systems*, Addison-Wesley, 1975.

2.9 Denning, P. J., 'Virtual memory', *ACM Computing Surveys*, **2**, 3, 153, 1970.

2.10 Diebold Group, *Automatic data processing handbook*, McGraw-Hill, 1977.

2.11 Donovan, J. J., *Systems programming*, McGraw-Hill, 1970.

2.12 Eckert, J. P., 'Thoughts on the history of computing', *Computer*, **9**, 12, 58, 1976.

2.13 Finneran, T. R., and Henry, J. S., 'Structured analysis for database design', *Datamation*, **23**, 11, 99, 1977.

2.14 Fry, J. P., and Sibley, E. H., 'Evolution of database management systems', *ACM Computing Surveys*, **8**, 1, 7, 1976.

2.15 Gepner, H. L., 'User ratings of software packages', *Datamation*, **23**, 12, 117, 1977.

2.16 Gold, M. M., 'Time-sharing and batch processing: an experimental comparison of their values in a problem solving situation', *Communications of the ACM*, **12**, 5, 249, 1969.

2.17 Haidinger, T. P., and Richardson, D. P., *A manager's guide to computer time-sharing*, John Wiley and Sons, 1975.

2.18 Higman, B., *A comparative study of programming languages*, Macdonald/American Elsevier, 1967.

2.19 Hill, F. J., and Peterson, G. R., *Digital systems: Hardware organization and design*, John Wiley and Sons, 1973.

2.20 Hoare, C. A. R., and Perrott, R. H., *Operating systems techniques*, Academic Press, 1972.

2.21 Johnson, C. I., 'Interactive graphics in data processing: Principles of interactive systems', *IBM Systems Journal*, 7, 3, 147, 1968.

2.22 Jones, A. V., 'A user oriented database retrieval system', *IBM Systems Journal*, 16, 1, 4, 1977.

2.23 Knuth, D. E., *The art of computer programming, Vol. 3 Sorting and searching*, Addison-Wesley, 1973.

2.24 Knutsen, K. E., and Nolan, R. L., 'Assessing computer costs and benefits', *Journal of Systems Management*, 25, 2, 28, 1974.

2.25 Lewis, L. J., 'Service levels: A concept for the user and the computer centre', *IBM Systems Journal*, 15, 4, 328, 1976.

2.26 Lorin, H., *Sorting and sort systems*, Addison-Wesley, 1975.

2.27 Martin, J., *Design of real-time computer systems*, Prentice-Hall, 1967.

2.28 Martin, J., *Telecommunications and the computer*, Prentice-Hall, 1969.

2.29 Martin, J., *Computer and database organization*, Prentice-Hall, 1975.

2.30 Maynard, J., *Modular programming*, Auerbach Publishers, 1972.

2.31 McGee, W. C., 'The information management system IMS/VS Part 1: General structure and operation', *IBM Systems Journal*, 15, 4, 328, 1976.

2.32 Michaels, A. S., Mittman, B., and Carlson, C. R., 'A comparison of relational and CODASYL approaches to database management', *ACM Computing Surveys*, 9, 4, 273, 1977.

2.33 Miller, I. M., 'Computer graphics for decision making', *Harvard Business Review*, 47, 6, 121, 1969.

2.34 Miller, R. B., 'Response time in man–computer conversational transactions', *Proc. AFIPS Fall Joint Computer Conference*, 267, 1968.

2.35 Myers, W., 'Key developments in computer technology—A survey', *Computer*, 9, 11, 48, 1976.

2.36 Naur, P., *Concise survey of computer methods*, Studentlitteratur, Lund, 1974.

2.37 Ord-Smith, R. J., and Stephenson, J., *Computer simulation of continuous systems*, Cambridge University Press, 1975.

2.38 Palmer, I. R., *Database systems: a practical reference*, C.A.C.I., London, 1975.

2.39 Randell, B., *The origins of digital computers—selected papers*, Springer Verlag, 1973.

2.40 Rosen, S., 'Electronic computers: a historical survey', *ACM Computing surveys*, 1, 1, 7, 1969.

2.41 Sammett, J. E., *Programming languages: History and fundamentals*, Prentice-Hall, 1969.

2.42 Swanson, E. B., 'Management information systems: Appreciation and involvement', *Management Science*, 21, 2, 178, 1974.

2.43 Taylor, R. W., and Frank, R. L., 'CODASYL database management systems', *ACM Computing Surveys*, 8, 1, 43, 1976.

2.44 Tucker, A. B., *Programming languages*, McGraw-Hill, 1977.

2.45 Veinott, C. G., 'Programming decision tables in Fortran, Cobol, or Algol', *Communications of the ACM*, 9, 1, 31, 1966.

2.46 Watson, R. W., *Time-sharing system design concepts*, McGraw-Hill, 1970.

2.47 Wells, M., *Computing systems hardware*, Cambridge University Press, 1976.

2.48 Weinberg, G. M., *The psychology of computer programming*, Van Nostrand, 1971.
2.49 Wilkes, M. V., *Time-sharing computer systems*, Macdonald/American Elsevier, 1972.
2.50 Withington, F. G., 'Five generations of computers', *Harvard Business Review*, **52**, 4, 99, 1974.

3 Computer applications and techniques in production planning and control systems.

3.1 Berry, W. L., 'Lot sizing procedures for requirements planning systems: A framework for analysis', *Production and Inventory Management*, **13**, 2, 19, 1972.
3.2 Box, G. E. P., and Jenkins, G. M., *Time series analysis, forecasting and control*, Holden Day, 1970.
3.3 Brennan, J., *Operational research in industrial systems*, English Universities Press, 1972.
3.4 Brown, R. G., *Smoothing, forecasting and prediction of discrete time series*, Prentice-Hall, 1963.
3.5 Brown, R. G., *Decision rules for inventory management*, Holt, Rinehart and Winston, 1967.
3.6 Buffa, E. S., and Taubert, W. H., *Production–inventory systems: planning and control*, Richard D. Irwin, 1972.
3.7 Buffa, E. S. *Operations management: problems and models* John Wiley and Sons, 1972.
3.8 Buffa, E. S., *Modern production management*, John Wiley and Sons, 1973.
3.9 Burbridge, J. L., *Production planning*, Heinemann, 1971.
3.10 Carter, L. R., and Huzan, E., *A practical approach to computer simulation in business*, George Allen and Unwin, 1973.
3.11 Chambers, J. C., Mullick, S. K., and Smith, D. D., *An executive's guide to forecasting*, John Wiley and Sons, 1974.
3.12 Cobham, A., 'Priority assignment in waiting line problems', *Operations Research*, **2**, 1, 76, 1954.
3.13 Conway, R. W., Johnson, B. M., and Maxwell, W. L., 'An experimental investigation of priority dispatching', *Journal of Industrial Engineering*, **11**, 3, 221, 1960.
3.14 Conway, R. W., 'Priority dispatching and work-in-process inventory in a job shop', *Journal of Industrial Engineering*, **16**, 2, 123, 1965.
3.15 Conway, R. W., 'Priority dispatching and job lateness in a job shop', *Journal of Industrial Engineering*, **16**, 4, 228, 1965.
3.16 Conway, R. W., Maxwell, W. L., and Miller, L. W., *Theory of scheduling*, Addison-Wesley, 1967.
3.17 Corke, D. K., *Production control in engineering*, Edward Arnold, 1977.
3.18 Coutie, G. A., Davies, O. L., Hossell, C. H., Millar, D. W. G. P., and Morrell, A. J. A., *Short term forecasting: ICI Monograph No. 2*, Oliver and Boyd, 1964.
3.19 Dematteis, J. J., 'An economic lot sizing technique, Part I, The part period algorithm', *IBM Systems Journal*, **7**, 1, 30, 1968.

3.20 Draper, N. R., and Smith H., *Applied regression analysis*, John Wiley and Sons, 1968.
3.21 Dzielinski. B. P., Baker, C. T., and Manne, A. S., 'Simulation tests of lot size programming', *Management Science*, **9**, 2, 229, 1963.
3.22 Eilon, S., *Elements of production planning and control*, Macmillan, 1962.
3.23 Forster, A. J., and Hsu, E. Y. P., 'Material requirements planning—The first line between success and failure', Proc. 19th annual conference of the American production and inventory control society, Atlanta, Georgia, USA, 421, 1976.
3.24 Gavett, J. W., *Production and operations management*, Harcourt, Brace and World, 1968.
3.25 Green, J. H., *Production and inventory control handbook*, McGraw-Hill, 1970.
3.26 Hadley, G., and Whitin, T. M., *Analysis of inventory systems*, Prentice-Hall, 1963.
3.27 Harrison, P. J., 'Short term sales forecasting' *Applied statistics*, **14**, 2, 102, 1965.
3.28 Hillier, F. S., and Lieberman, G. J., *Introduction to operations research*, Holden Day, 1974.
3.29 Johnson, R. A., Newell, W. T., and Vergin, R. C., *Operations management: A systems concept*, Houghton Mifflin, 1972.
3.30 King, J. R., *Production planning and control: An introduction to quantitative methods*, Pergamon, 1975.
3.31 Lock, D., *Industrial scheduling techniques*, Gower Press, 1971.
3.32 Lockyer, K. G., *Production control in practice*, Pitman, 1975.
3.33 Magee, J. F., and Boodman, D. M., *Production planning and inventory control*, McGraw-Hill, 1967.
3.34 Milne, T. E., *Business forecasting—A managerial approach*, Longman, 1975.
3.35 Mize, J. H., White, C. R., and Brooks, G. H., *Operations planning and control*, Prentice-Hall, 1971.
3.36 Morris, R. G., and Cubbin, M. J., 'Simulation of material handling methods', *Industrial Engineering*, **3**, 10, 44, 1971.
3.37 Morrell, J., *Management decisions and the role of forecasting*, Penguin, 1972.
3.38 Nicholson, T. A. J., and Pullen, R. D., *Computers in production management decisions*, Pitman, 1974.
3.39 O'Gorman, K. L., 'Capacity planning in the 1960's 1970's and 1980's', Proc. 19th annual conference of the American production and inventory control society, Atlanta, Georgia, USA, 289, 1976.
3.40 Orlicky, J. A., 'Net change material requirements planning', *IBM Systems Journal*, **12**, 1, 2, 1973.
3.41 Orlicky, J. A., *Material requirements planning*, McGraw-Hill, 1975.
3.42 Plossl, G. W., and Wight, O. W., *Production and inventory control principles and techniques*, Prentice-Hall, 1967.
3.43 Riggs, J. L., *Production systems: planning, analysis and control*, John Wiley and Sons, 1970.
3.44 Taha, H. A., and Skeith, R. W., 'The economic lot sizes in multi-stage production systems', *Transactions of the AIIE*, **2**, 2, 157, 1970.

3.45 Trigg, D. W., and Leach, A. G., 'Exponential smoothing with an adaptive response rate', *Operational Research Quarterly*, **18**, 1, 53, 1967.

3.46 Tocher, K. D., *The art of simulation*, English Universities Press, 1964.

3.47 Wagner, H. M., and Whitin, T. M., 'Dynamic version of the economic lot size model', *Management Science*, **5**, 1, 89, 1958.

3.48 Wight, O. W., *Production and inventory management in the computer age*, Cahners Books, 1974.

3.49 Wild, R., *The techniques of production managment*, Holt, Rinehart and Winston, 1971.

3.50 Winters, P. R., 'Forecasting sales by exponentially weighted moving averages', *Management Science*, **6**, 3, 324, 1960.

4 Computer systems implementation methodology

4.1 Arms, W. Y., Baker, J. E., and Pengelly, R. M., *A practical approach to computing*, John Wiley and Sons, 1976.

4.2 Aron, J. D., *The program development process*, Addison-Wesley, 1974.

4.3 Blackman, M., *The design of real-time applications*, John Wiley and Sons, 1975.

4.4 Boies, S. J., 'User behaviour on an interactive computer system' *IBM Systems Journal*, **12**, 1, 2, 1973.

4.5 Booth, G. M., *Functional analysis of information processing*, John Wiley and Sons, 1973.

4.6 Brown, P. J., 'Programming and documenting software projects', *ACM Computing Surveys*, **6**, 4, 213, 1974.

4.7 Buckle, J. K., *Managing software projects*, Macdonald/American Elsevier, 1977.

4.8 Coan, D. R. A., and Sharratt, J. R., *Programming standards 1: Documentation*, NCC Publications, 1976.

4.9 Coan D. R. A., and Sharratt, J. R., *Programming standards 2: Techniques*, NCC Publications, 1976.

4.10 Coan D. R. A., *Standards management*, NCC Publications, 1977.

4.11 'Computer Security: Backup and recovery methods', *EDP Analyzer*, **11**, 9, 1973.

4.12 Coleman, R. J., and Riley, M. J., *MIS management dimensions*, Holden Day, 1973.

4.13 Couger, J. D., 'Evolution of business systems analysis techniques', *ACM Computing Surveys*, **6**, 4, 213, 1974.

4.14 Couger, J. D., and Knapp, R. W., *System analysis techniques*, John Wiley and Sons, 1974.

4.15 Cuozzo, D. E., and Kurtz, J. F., 'Building a base for database: A management perspective', *Datamation*, **19**, 10, 71, 1973.

4.16 Davis, G. B., *Management information systems: Conceptual foundations, structure, and development*, McGraw-Hill, 1974.

4.17 Ein-Dor, P., 'A dynamic approach to selecting computers' *Datamation*, **23**, 6, 103, 1977.

4.18 Gibbons, T. K., *Integrity and recovery in computer systems*, NCC/LSE/Heyden, 1976.

4.19 Glein, G. A., *Electronic data processing systems and procedures*, Prentice-Hall, 1971.

4.20 Holden, G. K., *Production control packages and services*, NCC Publications, 1976

4.21 Howarth, R. J., and Lim, A. L., 'An approach to program documentation', *Computer Bulletin*, **13**, 8, 291, 1969.

4.22 Johnson, R. A., Kast, F. E., and Rosenzweig, J. E., *The theory and management of systems*, McGraw-Hill, 1967.

4.23 Kanter, J., *Management-oriented management information systems*, Prentice-Hall, 1977.

4.24 Kochhar, A. K., and Parnaby, J., 'The choice of computer systems for real-time production control', *Production Engineer*, **56**, 10, 29, 1977.

4.25 Kochhar, A. K., 'Distributed real-time data processing for manufacturing organizations', *IEEE Transactions on Engineering Management*, **EM 24**, 4, 119, 1977.

4.26 Lenven, K. V., 'Manufacturing systems justification—A technique', Proceedings 19th annual conference of the American production and inventory control society, Atlanta, Georgia, USA, 54, 1976.

4.27 Li, D. H., *Design and management of information systems*, Science Research Associates, 1972.

4.28 Lockyer, K. G., *An introduction to critical path analysis*, Pitman, 1967.

4.29 Lucas, H. C., 'A user-oriented approach to system design', Proc. ACM 1971 Annual Conference, ACM, New York, 325, 1971.

4.30 Malia, T. C., and Dickson, G. W., 'Management problems unique to on-line real-time systems', Proc. AFIPS 1970 Fall Joint Computer Conference, 569, 1970.

4.31 Martin, J., *Design of man–computer dialogues*, Prentice-Hall, 1973.

4.23 Maynard, J., *Computer programming management*, Butterworths, 1972.

4.33 McFarlan, F. W., 'Problems in planning the information system', *Harvard Business Review*, **49**, 2, 99, 1971.

4.34 Meadow, C. T., *Man-machine communication*, John Wiley and Sons, 1970.

4.35 Meister, D., *Behavioural foundations of system development*, John Wiley and Sons, 1976.

4.36 Metzger, P. W., *Managing a programming project*, Prentice-Hall, 1973.

4.37 Miyamoto, I., 'Software reliability in on-line real-time environment', Proc. International Conference on Reliable Software, Los Angeles, USA, 194, 1975.

4.38 Moder, J. J., and Phillips, C. R., *Project management with CPM and PERT*, Van Nostrand, 1970.

4.39 National Computing Centre, *Documenting systems (The user's view)*, NCC Publications, 1972.

4.40 National Computing Centre, *Systems documentation manual*, NCC Publications, 1975.

4.41 Rothery, B., Mullally, A., and Byrne, A., *The art of systems analysis*, Business Books, 1976.

4.42 Rothstein, M. F., *Guide to the design of real-time systems*, John Wiley and Sons, 1970.

4.43 Shaw, J. C., and Atkins, W., *Managing computer systems projects*, McGraw-Hill, 1970.

4.44 Taggart, W. M., and Tharp, M. O., 'A survey of information requirements analysis techniques', *ACM Computing Surveys*, **9**, 4, 273, 1977.

4.45 Tassel, D. V., *Program style, design, efficiency, debugging, and testing*, Prentice-Hall, 1974.

4.46 Tebbs, D., and Collins, G., *Real-time systems management and design*, McGraw-Hill, 1977.

4.47 Timmreck, E. M., 'Computer selection methodology', *ACM Computing Surveys*, **5**, 4, 199, 1973.

4.48 Tricker, R. I., *Management information and control systems*, John Wiley and Sons, 1976. ·

4.49 Walsh D. A., *A guide for software documentation*, McGraw-Hill, 1969.

4.50 Waters, S. J., *Introduction to computer systems design*, NCC Publications, 1974.

4.51 Weinwurm, G. F., *On the management of computer programming*, Auerbach Publishers, 1970.

4.52 Wong, K. K., *Computerguide 5: Concepts on program testing*, NCC Publications, 1972.

4.53 Wong K. K., and Joyce, M. A., *Factfinder 8: program testing aids*, NCC Publications. 1972.

4.54 Yohe, J. M., 'An overview of programming practices', *ACM Computing Surveys*, **6**, 4, 221, 1974.

4.55 Young, R. B., 'The computer and the contract', *Datamation*, **17**, 11, 22, 1971.

4.56 Yourdon, E., *Design of on-line computer systems*, Prentice-Hall, 1972.

4.57 Zani, W. M., 'Blueprint for MIS', *Harvard Business Review*, **48**, 6, 95, 1970.

Glossary

ABC Analysis: Method of grouping inventory items into different categories based on their monetary value and usage.

Access Time: The total time required to reach a particular record on the storage medium.

Addressing Scheme: Technique for specifying the location of an instruction or record.

Algorithm: Rule for solving a mathematical problem in a number of steps.

Arithmetic and Logic Unit: The element of the central processing unit (CPU) which carries out the necessary arithmetical and logical manipulations of data.

Artificial Intelligence: The application of computers for performing very high level tasks, such as learning and decision making, normally performed by human beings.

Assembly Language: A low level symbolic language in which operations are normally represented by easily remembered mnemonics. An assembler program is required to translate the assembly language instructions into the machine code executed on the computer.

Asynchronous Transmission: Method of communication involving the transmission of data one at a time, Generally used for slow speed terminals operated by human beings.

Background/Foreground Mode: A mode of system operation that provides for the execution of two programs. The foreground program has priority, but whenever sufficient resources are available the background program is allowed to execute.

Backward Scheduling: Capacity planning principle based on the assumption that all the items will be produced by the due date. The start date is calculated by taking account of the lead time for all the operations necessary to manufacture the item.

Batch Processing: Mode of computer operation in which the transactions taking place at random intervals are collected prior to their processing at a predefined time to produce an updated file. Alternatively, batch processing refers to a computer system in which the competing jobs are queued for execution.

Bit: A digit with possible values of 0 and 1. Also referred to as a binary digit.

Buffer: Storage medium for temporarily holding data to be transferred from one unit to another.

Byte: Collection of 8 bits capable of representing an alphanumeric or special character.

Capacity Planning: Process of planning the machine and manpower capacity required to satisfy the current and projected requirements for finished products and spare parts.

Central Control Unit: A unit that exercises control over the operations of the computer.

Central Processing Unit (CPU):   Collective term for describing the three main elements of a digital computer. These are (a) Arithmetic and Logic unit, (b) Central Control Unit, (c) Main Memory.

Chained file: A file in which pointers are used to link together records containing common identifying fields of information.

Compiler: A system program for translating instructions in high level languages, such as Cobol or Fortran, into machine code executed on the computer.

Complement: A number derived from an existing number in accordance with a specified rule. In many computer systems the complement of a number is used to store negative values.

Computer Output on Microfilm (COM):   An information output medium increasingly used in place of line printers. Information can be recorded on microfilm at very high speeds.

Computer Software Packages: A program or set of programs used to carry out specific functions, such as inventory recording or statistical forecasting, in a predefined manner. The availability of such packages, provided they are suitable for the application under consideration, reduces the need for the development of expensive in-house systems.

Database: An integrated collection of data in which the natural and logical relationships between various data items, for example assemblies and components, are automatically represented thereby improving the flow and accuracy of information.

Debugging: Process of detecting and correcting the errors in a computer program.

Demand filter: A filter for checking that the current demand data to be used for computing the forecasts of future demand does not exhibit a large random disturbance.

Direct Access: Data storage method that makes it possible to access and retrieve the records in a random manner. Magnetic discs, drums, and main memories are examples of direct or random access storage media.

Disc Directory: Table for recording the location of individual files stored on the disc.

Exponential Smoothing: Technique whereby the weights attached to the current and historical data values decrease in an exponential manner. The actual weights are altered by modifying the value of the smoothing factor alpha (α).

Field: Smallest unit of data/information storage. In an inventory control system, for example typical fields would contain data such as part number, part description, quantity on hand, quantity on order, etc.

Forward Scheduling: Capacity planning technique in which the batches of work are loaded onto appropriate work centres, taking into account the processing, queuing and transit times for each operation, beginning with the start of the first operation on a specified future date or the current date.

High Level Language: A language for writing the computer program instructions in an easy to understand notation which does not require any knowledge of the machine code actually executed, or the computer architecture. Cobol and Fortran are examples of such high level languages. Compliers are used for translating the high level language instructions into machine code.

Index Register: A storage device used to store the address of some locations within the main memory.

Instruction Register: A storage device in which the address of the current instruction is stored.

Instruction Set: Group of machine code instructions implemented on a particular machine.

Interactive System: A system that allows the user to be in direct and meaningful contact with the computer via terminals such as visual display units, or character printers.

Interpretive Language: Programming language in which each source program statement is analysed and translated into object code every time it is necessary to execute that particular statement.

Interrupt: A feature frequently used in modern computer systems to draw the attention of the processor to the occurrence of particular events.

Job Control Language: A language for describing the resource requirements, for example CPU time or main memory, of a particular task (or program).

K: An abbreviation for 1024 (in computer terminology).

Low level Language: A programming language in which each programmed instruction has a corresponding machine code instruction.

Machine Code: The instructions directly executed on the computer.

Main Memory: Rapidly accessed internal memory of the computer in which the instructions for current use are stored.

Management Information System: A system intended to help the management plan and control the operations of an organization, i.e. aid in the decision making tasks.

Mark Sense Readers: Devices used to read the information, recorded on a sheet of paper, in the form of a series of marks.

Mass Storage Devices: Units, such as magnetic discs, capable of storing large volumes of data required in a computer installation.

Mean Absolute Deviation: Average of the absolute differences between the forecast and actual demand values.

Memory Management: Hardware or software techniques for allocating main memory space to the pages or segments of competing programs.

Micro-computer: Small computers, comprising single chip LSI (large scale integration) processors, suitable for dedicated applications.

Microprogramming: Hardware technique for the execution of a large number of interrelated micro-instructions. Useful for modifying the instruction set of a computer.

Monte Carlo Technique: A method, involving the use of random numbers, frequently employed in computer simulation applications.

Multiplexer: Hardware unit for connecting a number of slow speed terminals to the computer via a single high speed line.

Multi-programming: A technique for concurrently executing more than one program on the computer, thereby improving the utilization of the available resources.

Netting: Process of performing gross to net calculations, for finished products and lower level components, by taking account of the available inventory items.

Offsetting: Calculation of the date on which the procedures for procuring the net requirements must be started off.

On-line System: In such systems the operators use terminals, for example visual display units, to input data to or receive information directly from the computer system.

Operating System: A set of programs which define the control rules and proceedures used for the allocation of computer resources, and the supervision of the running of the individual application programs.

Operations Scheduling: Preparation of a realistic working schedule by establishing the expected start and completion times for individual manufacturing operations.

Optical Character Recognition: Technique for optically recognizing, at very high speeds, the characters printed by hand or machine on a piece of paper. The recognized characters are then translated into formats suitable for computer processing.

Overlaying: Technique for loading different sections of an executing program into the same area of the main memory.

Parity check: Method of ensuring that the data recorded on or read from the recording media, such as magnetic or paper tape, are correct.

Parts Explosion: Determination of components and raw materials required to produce a specified item.

PERT/CPM: Management techniques for planning and controlling large scale projects. The various activities and their durations are represented in the form of a network, and the critical path determined.

Read Only Memory (ROM):   Memory whose contents are fixed. The cycle time of ROM is less than that of random access memory (RAM).

Real-Time System: A system in which the transactions data are captured at source followed by immediate updating of the affected records. Such systems always provide up to date information to the users, via v.d.u.'s etc., thereby making it possible to control the environment in which the system is operating.

Response Time: Elapsed time between the transmission, from a terminal, of an enquiry to the computer system, and the receipt of a reply (or acknowledgement in some cases).

Shared Peripherals: Devices, such as line printers or discs, available for concurrent use by a number of programs.

Spooling: Technique for handling slow speed input/output terminals with the objective of increasing computer throughput.

Synchronous Transmission: Data transmission method in which the information is transmitted at a fixed rate in the form of a continuous string of a number of characters.

Time-sharing Computer System: Mode of computer operation that allows two or more users to run their programs simultaneously on the computer.

Uni-programming System: A computer system in which a program is executed until completed, or aborted due to the existence of errors.

Utility Programs: Programs, usually provided by the computer manufacturer, for performing frequently encountered tasks such as the interchange of information between different peripherals, and sort/merge operations.

Virtual Memory System: In such systems the allocation of the main memory to the different sections of currently active programs is carried out automatically using hardware/software techniques.

Index